Formal Modeling in Social Science

A formal model in the social sciences serves to build explanations—when it structures the reasoning underlying a theoretical argument, opens venues for controlled experimentation, and perhaps leads to hypotheses. Yet even more importantly, models evaluate theory, build theory, and enhance conjectures. *Formal Modeling in Social Science* explores fundamental considerations of epistemology and methodology and also discusses the varied helpful roles of formal models.

The authors integrate the exposition of the epistemology and the methodology of modeling and argue that these two aspects reinforce each other. They illustrate the process of designing an original model suited to the puzzle at hand, using multiple methods in diverse substantive areas of inquiry. The authors also emphasize the crucial, though underappreciated, role of a narrative in the progression from theory to model.

Transparency of assumptions and steps in a model means that any analyst will reach equivalent predictions whenever she replicates the argument. Hence, models enable theoretical replication, essential in the accumulation of knowledge.

Formal Modeling in Social Science speaks to scholars in different career stages and disciplines and with varying expertise in modeling.

Carol Mershon is Professor of Politics at the University of Virginia.

Olga Shvetsova is Professor of Political Science and Economics at Binghamton University.

FORMAL MODELING IN SOCIAL SCIENCE

Carol Mershon and Olga Shvetsova

University of Michigan Press
Ann Arbor

Copyright © 2019 by Carol Mershon and Olga Shvetsova
Cover photo copyright © 2019 by Gregory Filippov

Published in the United States of America by the University of Michigan Press
Manufactured in the United States of America
Printed on acid-free paper

A CIP catalog record for this book is available from the British Library.
Library of Congress Cataloging-in-Publication data has been applied for.

First published September 2019

ISBN: 978-0-472-07423-5 (Hardcover : alk paper)
ISBN: 978-0-472-05423-7 (Paper : alk paper)
ISBN: 978-0-472-12586-9 (ebook)

To Norman Schofield, always a friend

Contents

Digital materials related to this title can be found on
the Fulcrum platform via the following citable URL:
https://doi.org/10.3998/mpub.8811571

Tables

Figures

Acknowledgments

Work on this book differed from our previous projects in several ways, all of which involved interactions with many people. First, *Formal Modeling* synthesizes our professional training and research experience in political economy with our decades of teaching frustrations, when formal modeling is lumped together and judged at par with statistical analysis or seen as a mere embellishment to the process of discovery. Second, because we believe that the arguments in this book reflect not only our thinking but also collective practice in the discipline, we actively sought the input, preferably critical, of our colleagues. We were ready to hear divergent views, defend our position, and sharpen the argument. In light of that, we reached out to many scholars, which makes the final product a perhaps broader collective enterprise than is usually the case. Third, rising young scholars in political economy contributed by coauthoring chapters 5, 6, and 7 with us, and thus this project also became collective in the direct sense of the word.

All of these features made the journey immensely exciting and intellectually stimulating. We loved writing this book, and we loved interacting with many amazing people along the way. First and foremost, the brilliant readers commissioned by the University of Michigan Press internalized our approach and engaged with us fully. They should know how deeply we appreciate their contribution, and we hope that they can see that we have listened and acted. We are grateful indeed.

Colleagues at Binghamton and the University of Virginia gave generously of their time, attention, and ideas. We are very grateful to those who were willing to assign the book's early drafts in their PhD seminars. Their students furnished us with comments and we thank them as well. Friends and colleagues around the world read and commented on the book.

Here we list in alphabetical order those who had a role in our work on the book: Enriqueta Aragonès, Raj Arunachalam, Andrew Bennett, Hayley Black, Allison Bugenis, Tiziana Caponio, Margarita Estévez-Abe, Mikhail Filippov, Andrew Foote, Alena Gericke, Esther Hauk, William Heller, Charles Holt, Roya Izadi, David Leblang, Gilat Levy, Anne Meng, Syed Rashid Munir, Vhonani Netshandama, George Ofosu, Socorro Puy, Gregory Robinson, Chiara Saraceno, Katri Sieberg, Evgeny Sedashev, Didem Seyis, Stergios Skaperdas, Christine Sylvester, David Waldner, Denise Walsh, and Huei-Jyun Ye.

Many thanks to our research assistants for their diligence and good cheer: Simranjit Bhatia, Gabriella Gladney, Emily Laurore, Shiran Ren, Renato Sepulveda, and Jennifer Simons. We thank them all, and especially Shiran Ren, with whose assistance we pulled the book over the finish line.

Three colleagues joined us as chapter co-authors: Julie VanDusky-Allen (Boise State University), chapter 5; Benjamin Farrer (Knox College), chapter 6; and Andrei Zhirnov (Binghamton University), chapter 7. The partnerships we enjoyed with them have enriched the book throughout.

We acknowledge institutional and financial support from our home universities. Additional financial support came from the Collegio Carlo Alberto, Workshop in Political Economy in Falset 2018, European Research Council, MOVE (Markets, Organizations, and Votes in Economics), Institut d'Anàlisi Econòmica, and CSIC (Consejo Superior de Investigaciones Científicas). At the University of Virginia, an Arts, Humanities, and Social Sciences Research Award enabled us to visit each other, and the Hugh S. and Winifred Cumming Chair research funds supported copyright costs and the work of our indexer.

We also owe debts of gratitude to our supporters at the University of Michigan Press. We have benefited from the editorial expertise of, initially, Melody Herr and Mary Francis, and subsequently, Elizabeth Demers and Danielle Coty. Mary Hashman managed the production process with efficiency and a most appreciated mixture of clarity and flexibility. We thank Heidi Dailey and Gregory Filippov for the cover art. Our thanks as well go to those behind the scenes whose efforts and professionalism contributed to producing the book now between two covers.

Last, and most importantly, we give heartfelt thanks to our family and friends for their support and tolerance. Our deep dives into the project required heroic levels of love and patience on their part, in which they did not waver.

With so many people to thank, we alone assume full responsibility for any remaining shortcomings in the pages that follow.

We dedicate this book to Norman Schofield, currently Dr. William Taussig Professor of Political Economy at Washington University in St. Louis, who played and continues to play a foundational role in the development of formal theory and in political economy. We are grateful to know him as our friend.

ONE

Introduction

A formal model in the social sciences serves to both build and evaluate explanations as it structures the reasoning underlying a theoretical argument, tests an argument's logic, opens venues for controlled experimentation, and perhaps leads to hypotheses. The capacity of models to perform these roles makes them attractive to many analysts, and may have motivated some readers to pick up this book in the first place. This book addresses these roles and goes further to explore more fundamental considerations of the epistemology and methodology of formal modeling in social research. Whereas scholars often file modeling under the rubric of methodology, modeling also has an epistemological place.

Modeling clears a path to knowledge, playing an important part in how we know what we know. Formal models, the book demonstrates, *test theory*, *build theory*, and *enhance conjectures*. Models also make transparent the assumptions and steps in a scholar's theoretical argument, which means that other analysts can reproduce that argument and reach equivalent conjectures; that is, models *enable theoretical replication*. This role in theoretical replication underpins the role of models in *enabling accumulation of knowledge*.

The discussion of epistemology in chapter 2 buttresses the exposition of formal methodology, to which chapter 3 is devoted. These early chapters create the platform from which to design a variety of formal models in chapters 4 through 7—using multiple methods and reaching into a range of substantive areas of inquiry, one or a few of which may strike the reader as surprising. To wit, a traditional chief's decision to inaugurate an award ceremony in his local community, as reported in a single 2014 South African news story, inspires three research queries and three formal models (as in chapter 4). Our epistemological approach imposes a certain order: it compels us to move from a real-world news story to multiple research queries and then to multiple models, so that the "how to" and the "how do we know" aspects of formal modeling become interrelated and mutually reinforcing.[1]

This introductory chapter begins by spelling out a fundamental clarification. Throughout, when we refer to *designing* a formal model, this means building the model from scratch, based on the theoretical and substantive material at hand. Next, we preview our argument about the place of formal models in research design. The second section considers models as methodological tools. The third section offers an initial look at the dynamics among stories, narratives, theories, and formal models. The fourth section describes the book's disciplinary scope—within and beyond political science. The map of the book concludes.

1.1. Designing Formal Models

The epistemological argument here requires that formal models be *designed*. Chapters 1 through 3 introduce readers to the concept, principles, and practice of designing formal models, not simply applying them. The book uses the phrase "applying a model" to refer to the practice of taking a preexisting model off the shelf and using it to analyze a given subject. For instance, applications abound of the renowned Prisoners' Dilemma. The problem with applications, however, is that scholars often strain to fit their chosen subject within the confines of one out of a fixed set of formal constructs. As they do so, they bend their theoretical arguments to conform to the standard intuitions of "ready-made" models. In designing a model, instead, an analyst constructs a formal model afresh, so that it focuses on a particular set of interactions and it answers a particular question. Consider as an example one of the models in chapter 4, designed specifically in order to unravel the puzzle: Why would a South African chief, who expects to reign for life, bother to institute rewards for members of the community he rules? The researcher attains theoretical precision by designing a model specific to the query she poses: this is why the book emphasizes design rather than application.

1.1.1. The Place of Formal Models in Research Design

The message here about formal models in research design is twofold. First, misperceptions and confusion exist about the impact of formal models on subsequent research design. Formal models on occasion provoke strong, conflicting, and even antagonistic reactions because they are so powerful, on the one hand, and yet, on the other, are seen by some scholars as clashing with principles of research design. Yet a researcher using a formal model need not be forced into a straitjacket as she fashions a research design and contemplates whether any new hypotheses are generated by the model. Granted, political scientists most often deploy statistical estimations as their empirical strategy subsequent to a formal model—so much so that the highly productive EITM (empirical implications of theoretical models) program connotes to many researchers the use of statistical estimations (see Granato and Scioli 2004 for a classic statement). But, in fact, scholars have pursued a range of alternative research designs to probe the empirical relevance of their formal models, including analytic narratives, qualitative research, and mixed-method research (e.g., Aldrich 2011 [1995]; Heller 2002; Heller and Mershon 2008; Kalyvas 1996; Lewin 1988; Meng 2018a, 2018b; Mershon and Shvetsova 2011, 2013a, 2013b, 2014; Miller and Schofield 2003; Schofield and Sened 2006). The message on this point must be one of flexibility: though formal models can be evaluated empirically, they do not have to be. Furthermore, whenever any such evaluation is attempted, it can proceed in multiple ways besides hypothesis-testing.

Empirical appraisals of formal models lie beyond the book's scope. One of its chief goals, however, is to shed light on the relative places of theory, models, and data in the overall epistemology of social research. To achieve that aim, chapters 1 through 3 unpack the linkages among theory, models, and evidence and show how models *test*, *build*, and *replicate* theory, as well as enable *knowledge accumulation*. Chapters 4 through 7 illustrate how models can yield *propositions*, whether testable or nontestable, in whatever substantive area models are used. The book argues and demonstrates that formal modeling is integral to research design in the study of human behavior.

The second message is that models anchor theoretical contributions to social scientific knowledge through their capacity to capture an argument's separate moving parts and strictly enforce the explicit assumptions about how those moving parts connect. In these ways, models assess the validity and internal consistency of theoretical arguments. That is, models *test theory*. Models make different predictions for different parameter values—for different assumptions.[2] Thus, models *build* theory. They refine the agenda for empirical inquiry as well (cf., e.g., Bates et al. 2000a; 2000b). In so doing, models *enhance conjectures*. Moreover, a formal model is theoretically distinct in that nothing in it is ad hoc: it establishes a tight connection from one logical link to another in its theoretical chain. Set apart by this distinctive feature, a model permits analysts to take a separate step in research design, one that does not resemble or supplement other steps. Formal modeling constitutes an epistemological step positioned before or alongside the stage where we (may or may not) generate testable propositions. Formal models are neither decorative nor illustrative, but have a fundamental place in research design in political and social sciences, that of enabling theoretical replication.

1.1.2. Formal Models and Theoretical Replication

Consider a thought experiment. Imagine that we asked a group of scholars how formal models gained their current popularity in the study of human behavior. Their answers would probably range all the way from appreciation for models' mathematical aesthetics to succumbing to peer pressure to include "some game theory." The most common answer, however, would likely invoke the general notion that models bring substantial value added to scholarship. If we then probed into specifics, we would hear what might be termed the "standard story" on the contributions of models to research design. Models serve as logical tests of theories in the sense that they force adherence to a strict structure of reasoning, and by doing so expose any internal inconsistencies that otherwise might escape notice. Some scholars might also add that models afford insights into new, more refined, testable propositions.

Yet the diverse readings of the purposes of modeling raise a question of models' almost excessive versatility. Is it really possible that formal models serve as tools for, at the same time, logical testing of theories, hypothesizing about causal linkages, and even supplying deterministic accounts of known processes? Or do the multiple roles attributed to formal models imply that analysts are confused, and as a result the current use of models is inconsistent, or, even worse, arbitrary? Granted, such concerns are legitimate. Yet we invite readers to set them aside—for now. Any given tool can be more or less appropriate to a task at hand. Think of house keys applied to the task of opening parcels: not utterly unreliable, yet not ideal either. The reader who considers the current use of formal models in social science as at times misdirected or even unnecessary will find no argument from us; but neither will we specifically focus on any such misuse. We focus instead on what we believe to be models' good and proper place in the overall epistemological schema, and the many purposes that they may serve albeit not always at the same time.

This thought experiment indicates, in essence, that researchers who use models have confidence in their argument as well as in their ability to convince others of its validity. This is why they turn to formal models. The attribute we named is that the argument is replicable. For now, observe that formal models serve as a platform for theoretical replication by assuring the unique logicality of each step in the analyst's reasoning, given explicitly stated theoretical and parametric assumptions. To amplify, for an argument to be *theoretically replicable*, it must be based on a set of explicitly identifiable assumptions and develop via clear logical steps (i.e., satisfy logicality all the way from assumptions to conjectures).[3] If any other analyst were to start with the same assumptions, she must be able to derive from those assumptions equivalent conjectures, just as with empirical replication she must be able to obtain the same estimates from the same data. Clarity and transparency, regardless of the level of mathematical complexity, go to the heart of theoretical replication. The replication standard, as applied to formal models, holds because any fellow modeler can replicate the results of a model, as the path leading to results is clear and transparent. Since formal models meet the replication standard, the theories that they reflect do so as well.

Because formal models permit theoretical replication, they can be used and reused repeatedly. Even when designing a new model, a researcher can treat past models' predictions as known and incorporate those predictions into the new model. To illustrate, chapter 5 uses the Prisoners' Dilemma as a ready-made, trusted baseline, and moves beyond that renowned model to fashion three original models from the same real-world news story about fishing industry regulation. Chapter 5 thus exploits accumulated knowledge and in turn contributes to the stock of knowledge, furthering the *accumulation of knowledge*.

At this juncture, the reader may object: Can an analyst reason in prose alone, without formalizing? Yes, and many analysts do just this. Yet, contrary to what some readers might think, the construction of a prose-only argument is often much more complicated than the building of a formal model. Granted, when there are few steps in an argument, and the steps are straightforward, nonformal reasoning may seem easy, and may seem easy to "replicate," in the sense that another researcher should arrive at the same conclusions from the same premises. The shorter the logical bridge, the clearer the connection between premises and conclusions, and the easier it is to follow and evaluate nonformal arguments. Yet when the logical bridge is lengthy, a prose-only argument is accurate only for some specific conditions, that is, given additional nonobvious assumptions. In a related vein, consider that prose reasoning is prone to unobserved logical errors due to hidden or omitted premises. For example, step five in the prose reasoning cannot be readily checked for connection and consistency against step two; just that sort of check is possible and even simple when the reasoning is expressed formally. Hence, prose-only reasoning, though it can be persuasive, yields itself to quoting in lieu of replication.

Or look at this from a different angle: prose is linear, one-dimensional, requiring pieces of argumentation to be positioned in a chain. A formal model can be constructed in multiple dimensions, and thus presents the argument much more intuitively insofar as that argument involves multiple simultaneously moving parts, as most interesting arguments do.

1.2. Models as Methodological Tools

Applied formal theory in social science is a very specific methodology, even though it is often presented as an art to students—and to scholars at any stage of their careers but new to the craft—who are invited to apply it in their own research. The "art" of formal theory is often confusing to neophytes in its use. Approaching a formal model as a methodology helps cut through the confusion, as shown throughout this book. At base, formal theorizing consists of a methodological toolbox that enables the analyst to move from a set of specific stated premises to a set of equally specific conclusions, some of which may correspond to clear testable propositions.

Many scholars today are accustomed to viewing approaches such as experiments, statistical models, and qualitative case studies as components of their methodological toolbox. All these tools have in common that the analyst uses them after she advances her hypotheses. Formal models equally are methodological tools, but they sharpen the implications of our theoretical arguments *before* or even without the hypothesis stage.

As methodological tools, models are used for a purpose, not as ends in themselves. Models do not precede or exist independently of the questions asked. That is, they do not characterize the world around us in an objective way. Granted, we can identify and use criteria to help us judge a given formal model or a given statistical specification as more or less suited to studying a phenomenon. Yet a "perfect" model or specification for the phenomenon can only be the phenomenon itself. For example, a "perfect" model of a human body would *be* a human body, because anything else would depart from the original. Hence "the right model" of a particular situation, phenomenon, or behavior does not exist. Here a parallel with statistical models is perhaps surprisingly appropriate, as there are no uniquely "right" statistical specifications for explaining any sort of observed variation in the data.

Nonetheless, neither formal nor statistical models are arbitrarily put together. When different scholars depart from the same point but end up with different models, it does not mean that they have built their models in ad hoc ways. They might have started together and have diverged from each other before they reached the modeling stage. Such divergence is a fact of the scholarly enterprise: formal models should not be blamed for it.

Theory provides the source of divergence. A minimalist definition of a *theory* is a set of assumptions from which a number of propositions follow logically, as just indicated. A formal *model*, as a manifestation of theory, proceeds by some form of mathematical reasoning from an explicit, exhaustive set of assumptions. Therefore, theoretical divergence precedes modeling and begins with the difference in the questions being asked and answered, questions that give purpose to the inquiry. Any answer requires a question in order to be seen as an answer, and both the phenomenon being questioned as well as the form that the question takes are embedded from the outset in a theoretical approach. To many analysts, it is the question that, once posited, guides the theoretical focus. Others would contend that the theoretical argument, that is, the answer, comes first in scholarship. In light of this contrast, the issue arises: Which has primacy, the question or its theoretical answer? In stressing the role of narratives as a stage that at least implicitly precedes modeling, we procedurally equate these otherwise contrasting worldviews—from the narrative stage on. If the analyst has some (implicit) narrative in mind before fashioning the model, that means that both the theory and the question are in place prior to formalization.

1.3. Theory, Models, and Evidence: A First Look

This chapter sketches what the entire book explores: the relationships between and among theories, models, and data. All students of human behavior who work with formal models speak in some way to evidence from the real world around them. Alongside this extreme statement, examples extend ad infinitum. The highly influential model of structure-induced equilibrium (SIE) reflects—while abstracting away from—the role played by powerful committees in the U.S. House (Shepsle 1979; Shepsle and Weingast 1981, 1987). The similar but distinct model advanced by Laver and Shepsle (1990, 1996, 1998; cf. Austen-Smith and Banks 1990) reflects and abstracts from the institutions of parliamentary democracy. In international relations, models of mutual assured destruction (MAD) distill the essence of Cold War confrontation (among many others, Powell 1987). Models enter everyday parlance, as do the games of Chicken and the Prisoners' Dilemma.

We obtain the substance for designing models from observing human behavior as captured in *stories* and *narratives*, and we use these two distinct terms in particular ways. While the general use of the term "story" captures a group of traits or a tale told, this we term a *narrative*. It is thus about narrative that Arendt (1970, 115) writes: "Storytelling reveals meaning without committing the error of defining it." The book uses the concept of a *story* to indicate a raw collection of facts, a minimally processed account of a real-world event.

Since we start with stories when modeling behavior, the first thing that occurs is framing questions—or recognizing puzzles. (As noted, chapters 4 through 7 use news stories.) That said, we are agnostic with regard to the existence of the converse possibility, where a scholar is invested in a theory and interprets the story through the prism of that theory. As prisms, theories refract reality and slice it into different narratives addressing different questions. Both starting with a question and starting with a favorite theory compel the next step, the one in which the original story is filtered into a *narrative*. The book defines a narrative differently than do philosophers, linguists, anthropologists, or sociologists.[4] Specifically, a narrative constitutes a subset of facts in the story, selected according to a theory-biased question. This means that the researcher-as-narrator recounts the event so as to introduce theoretical bias into a raw story. It is crucial to distinguish between minimally processed accounts of real-world events (stories, as labeled here) and accounts tinged with the researcher's theoretical biases and queries (here, narratives).[5]

Many scholars deploy narratives as hooks for their journal articles, devices to introduce their question, theory, research design, and so forth.[6] Such an opening narrative is not only stylistic but is inherently substantive: the narrative is shot through with assumptions that deserve just as much scrutiny and explicit acknowledgment as the assumptions entailed in a formal model. The narrative both is imbued with theoretical meaning and forms part of the research design, for it enables the researcher to abstract away from the infinite number of variables that appear in the story but are not invoked in the theoretical argument she builds.

The same set of facts, or story, can give rise to multiple analyses within different theoretical frameworks. At the stage of constructing their narratives from the same story, scholars already diverge from each other. Indeed, they do so when they select elements from the initial set of facts to include or omit in their narrative, aspects to highlight or underplay, implied questions to ask or answer, or actors to bring to the forefront or leave in the shadows. This process of constructing narratives out of the story is thus vital: the assumptions built into a narrative are inherited in the resulting formal model.

Thus, analysts who start with a question, like those who start with a theory, have both a real-world story and a narrative at least implicitly in mind. Even when a scholar seems to work from no obvious real-world point of departure, a narrative is still present; it can be imaginary and stylized. Even when a scholar omits the narrative stage in presenting her work, she still implicitly includes it in her thinking. That is, a student of human behavior always thinks in terms of narratives, whether or not she intends to formalize her reasoning. Likewise, even when the analyst does not explicitly reference a story, her implicit origins in factual material are reflected in the way she names the actors in her narrative (argument) and arranges the argument's moving parts.

Once more, all narratives are *biased* representations of stories, because they not only reduce events to an inevitable subset of evidence, but also originate, subjectively, with the analyst. The multiplicity of narratives reflects the fact that what is included and what is excluded at the recounting stage depends on the narrator's perspective. Different narrators—or the same narrator viewing an event from different angles—fashion multiple narratives. When narratives are closely tied to actual events, they are historical models that can also exist in the plural, as debates about interpretation of history reveal.

Throughout our exposition, our proposed *sequence* of story-narrative-model is transparent and transparently tracked. Delineating the relationship among stories, narratives, and models underscores the difference between a set of facts (a story) and the separate recounting of events (a narrative). The transition from the former to the latter stands as a meaningful, identifiable step in research design. After a scholar commits to having a formal model, this recounting stage becomes of obvious importance and is worthy of methodological scrutiny: the narrative exemplifies the workings of the theoretical mechanism envisioned by the scholar.

As a rendering of a story, a narrative makes the question under investigation into an obvious puzzle and also hints at the answer. Narratives serve these functions via a winnowing process, by filtering out nonrelevant items that are found in the story—actors, details of their behavior, even resulting outcomes—and by selectively extracting and highlighting the information relevant for both presenting the question and constructing the theory-consistent answer to it. While narratives are theoretically biased abstractions from the reality they reflect, they cannot and are not intended to reproduce that reality by themselves. We assert that narratives are first-order, nonformal models and treat them as such throughout. Likewise, formal models are theoretically biased abstractions. Models are intended to, and do, reconstruct reality in a stylized, biased, yet functional universe. With the stage of building narratives made explicit, it becomes clear that formal models depend on—that is, stem from—narratives. In this sense, formal models are second-order models: they are models of narratives.

When formal models are seen as extensions of narratives, it is to be expected that they distort or occasionally ignore reality. The statement just made may appear to concede to critics of modeling in the study of human behavior: Who at one time or another has not heard or uttered a critique that a model had little to do with reality? Yet as abstractions from reality, formal models are not any worse or better than written histories. We are accustomed, after all, to accepting the value of well-researched and well-written histories. Identifying this parallel between models and narratives is helpful for the purposes here: it underscores the epistemological value of models.

Because models are extensions of narratives, models are subjective by construction, since the narratives from which they are built are subjective (cf. Rubinstein 1991). Their subjectivity is rooted in the narrator's bias, which in turn comes from her preferred theoretical framework and consists in making implicit or explicit assumptions as she recounts her chosen story within that theoretical framework. Models carry these assumptions all the way through to their logical implications and along the way allow researchers to explore a variety of theoretical constraints and conditionalities on their arguments.

Both the narrative and the formal model represent the researcher's theoretical argument, all the while claiming to depict her real-world phenomenon of interest (or story). In a sense, then, scholars have a choice: to interpret whatever they study in the shape of a formal model or stop at a narrative. Yet this possibility of presenting to the reader just one out of the two begs the question: How are models and narratives related? For instance, how can a single narrative inspire multiple formal models? How does a model aid in evaluating the generic structure of interactions appearing in a given narrative? Chapters 2 and 3 probe these questions more deeply.

In chapters 4 through 7, we practice what we preach, taking as starting points ordinary, innocuous news stories that cover real-world events as information and without any apparent bias. To achieve this aim, our selection of the news items follows four criteria geared to minimize analytical bias in the stories. First and most simply, these news stories report on subject matter customarily of interest to some broad field in the social sciences. Second, the stories involve actors engaged in conflict. This might at first blush seem a criterion that would often be fulfilled but as it turns out is not. Third, these stories offer some sign of meaningful interaction, that is, interaction in which an individual actor is aware of other actors' actions and perceptions, actors are aware of their shared awareness, and actors behave so as to pursue advantage. Fourth, in these news stories the journalists report multiple points of view, quote multiple sources, and refrain from analytical interpretations or causal claims. This criterion ensures that whatever we select meets our definition of a story and is not in itself a narrative. This last criterion can hardly be satisfied in reporting momentous occasions, as readers and editors want not just facts but explanations and projections when, for example, electoral reversals or acts of terror occur. This criterion is also more easily met in some journalistic traditions than in others. To fulfill this criterion, we are best served by routine coverage of mundane events possibly even still in progress. To drive home that these stories rest on ordinary—literally, everyday—events, we search for stories on or near an arbitrary date, September 23, 2014.

Chapters 4 through 7 take these minimally processed accounts of events (or stories) found in the media to extract multiple narratives, and then convert them into formal models. The exposition in each of these chapters travels the entire distance from the most concrete to the most abstract representations of different interactions. The multiple journeys throughout each chapter repeatedly show models being built afresh—designed—to constitute theoretical generalizations of the specifics in stories and provide the basis for theoretical replication.

These journeys lay bare the process of discovery triggered by the starting question and theoretical approach. That process of discovery unfolds in analyzing a story and then a narrative within the preset confines of the chosen question and theory, whatever those may be. We have no goal of finding "the best" question or "the right" theory. We allow for a diversity of questions and theories, in an explicit acknowledgment that an analyst can adopt whatever theory and commit to it—before constructing a narrative and building a model. Moreover, as we develop each narrative in chapters

4 through 7, we characterize and then explain a class of phenomena rather than a specific observation from which we begin (cf. Carnap 1966).

1.4. Why This Book?

This book answers a set of complex questions about formal models, both epistemological and methodological in nature, some of which might seem to be at cross-purposes. In addressing these, the book shows that they do not clash but rather align. Scholarship that addresses the place of formal models in social research often focuses on defending their use. This leads to controversies, as various individual defenses fail to apply universally and also appear mutually contradictory. Models are defended either instrumentally for predictive capacity or realistically in terms of accuracy of assumptions (cf., e.g., MacDonald 2003).

This book focuses instead on the epistemological place of formal modeling. The *functions* of models, unlike their epistemological place, may be many and varied, differing across pieces of research. We discuss various functions of models in chapters 2 and 3, and in chapters 4 through 7 we witness models fulfilling some such functions. Useful functions of models are but added benefits compared to their epistemological role in the accumulation of knowledge.

1.4.1. Why Care about Models: Epistemology and Function

We test our theoretical arguments and come up with a better theory: this is the reason that models are essential for the accumulation of knowledge. The rest is a bonus.[7] As the book links the epistemology and methodology of formal modeling, it illustrates the place of modeling in the research design of a substantive scholar, and it shows how to construct models to suit a scholar's theory. It is the place in substantive research design where we care the most about models' functions, not just epistemology. As noted, it is widely accepted that formal models can generate hypotheses. This book moves beyond that commonplace and presents a unified exposition of how formal models test and build theory, and how models enhance hypotheses, permit theoretical replication, and enable knowledge accumulation.

Epistemologically, models *test the theories* from which they have sprung. A formal model forces the scholar to make all her assumptions explicit, and then identifies whether any assumptions are redundant or inconsistent. Models establish strict logicality within a theory, and confirm the logicality of an explanation. They can expand the set of potential predictions of an explanation, while confirming an explanation's original predictions. Models establish conditions in terms of parameter intervals, within which predictions hold. In all of these ways, then, a model allows the analyst to test her theoretical argument.

In doing so, models *build theory* by creating new theory from scratch or by improving extant theory. Models improve theory, in turn, by refining or extending it. Consider refinement of theory: struggling to maintain logicality of conclusions, the analyst may be compelled to adjust her assumptions, that is, to abandon her original theory in favor of amended theory. Modeling guides us through the process of refinement and also extension of theory. The book is organized so as to convey and illustrate this capacity of models to build theory.

Functionally, models can *enhance hypotheses*. Models might not only yield hypotheses but also improve hypotheses: observe the distinction. Models clarify the theory-specific set of dependent variables by establishing what does and does not follow from the theory at hand. By deriving the parametric bounds within which results exist, models generate hypotheses about empirical constraints on the existence and logicality of anticipated predictions. By deriving comparative statics on the dependent variable(s) with respect to specific parameters, models can lead to hypotheses about marginal effects and about global or local extreme values of the dependent variable conditionally on those parameters. By predicting complex multi-step and contingency-based sequences of individuals' actions, models can generate hypotheses about the presence of such complex behaviors, where the dependent variable can be defined as a sequence of contingency-based individual actions. A model allows an analyst to separate those outcomes that are idiosyncratic (pertinent to a specific theory on which a given model is based) from the outcomes that are logical consequences and thus qualify as true predictions of that theory.

Insofar as formal models enable the analyst to specify hypotheses, they can open up nonintuitive insights on the factual world. This point merits attention, for it is at times said that formal models merely generate predictions confirming what we already know. On the contrary, models can lead to predictions that are counterintuitive and thus otherwise would have eluded us. Furthermore, the implications yielded by models may not be testable, as when models predict inaction or a range of behaviors (cf. Clarke and Primo 2012). To clarify: hypotheses testing is not *the* goal of modeling, but it is sometimes the goal, and is often opportunistically attained.[8] Even when not leading to hypotheses, models are useful critical tools (cf. Hands 1991).

Models *enable theoretical replication*. As a mathematical object, a formal model establishes a structure of reasoning that is tight, solid, and strong. In a model, the process of building that structure of reasoning is clear, and each step in the process must be connected and consistent with all others—even though, as noted, a model may not be linear. Formal models emerge as an accountability device for an analyst, holding her to a standard of transparency in her theorizing that enables others to retrace every step in her reasoning, follow all steps through to her conclusions, and concur with her all along the way.[9] Thus, a formal model permits replication of theoretical arguments.

Models *enable knowledge accumulation*. Without fully repurposing extant formal models, scholars often benefit from incorporating existing, established models as building blocks in their original designs of new models as they seek to analyze particular empirical puzzles. This practice illustrates how the discipline engages in theoretical replication and advances accumulation of knowledge. This is not a conundrum. We have already previewed the treatment in chapter 5 of the Prisoners' Dilemma as a baseline for designing new models. We now add that, by the end of chapter 5, we discover that we in fact create a sequence of new models where we include prior models as building blocks in designing subsequent models. We also see that our models leave open the possibility of additional modeling, building up on what we do. While not undertaking that additional modeling ourselves, we spotlight what another analyst can borrow, both exploiting what we do and progressing further, thus advancing the accumulation of knowledge.

Chapters 4 through 7 put into action the design principles laid out in chapters 2 and 3. In doing so, they consistently follow the epistemological path in which we embed formal models. Even so, they intentionally diverge on formal modeling methods. Each of these chapters starts with a story, an empirical datum. The story has the raw materials for several narratives and several formal models. In satisfying the selection criteria, all stories feature people and conflict and no story implies a unique puzzle as the anchor for modeling. Which question will the modeler choose to investigate? What assumptions will she select? How will she hew to her assumptions as she develops her model? The answers to these questions yield diverse narratives and diverse models. The path from a single story to many models also demonstrates the rigor of modeling as a methodology: after the analyst has chosen the theoretical path in the construction of the narrative, she selects assumptions for each model and then must stick steadfastly to them and see where they take her.

1.4.2. Who This Book Is For

We speak to social scientists who incorporate formal modeling in their substantive research. In political science, many who are motivated by real-world puzzles and interested in evaluating theoretical explanations of those puzzles ponder whether and why including formal models is appropriate or even necessary. These questions are our main focus.

Social scientists may also find the book helpful in understanding and differentiating the variety of types of formal models. Chapters 4 through 7 furnish practical insights into how we fashion models as we seek to shed light on substantive research queries. Note that, though we expect readers to have a basic background in rational choice theory and game theory, we do not have any expectation beyond that. Both theoretical discussion and model design are accessible at that level.[10]

In what follows, we draw on literatures in economics, public policy, sociology, anthropology, and philosophy of science as well as political science. In turn, we claim relevance for our argument in the social scientific community. To wit, economists, sociologists, anthropologists, and social psychologists may have research interests in collective action. While some of these colleagues may already use formal modeling, others may not. All such readers share a concern with free riding (cf. Olson 1965). Chapter 4's formal treatment and substantive discussion of the fear of free riding in a South African traditional community speaks to these readers and others, while chapter 5 explicitly builds on the extant knowledge of the Prisoners' Dilemma as a representation of the Tragedy of the Commons. The multiple models of regulation in chapter 5 appeal to political economists trained (or training) in economics and public policy. Chapter 7's multiple models of abuse in a local jail speak to scholars in such fields as law and society, criminology, anthropology, race and politics, and public policy, whether they are newcomers to, or practitioners of, modeling. Furthermore, instructors of graduate courses in social science, including political science, may find the individual story-based chapters to be useful sources of illustrations of both the design of models and their suitability to substantive research queries. Finally, we hope to draw critical scrutiny from students of epistemology, on the subject of the place of formal reasoning in the acquisition of knowledge.

For all of our readers, we do not aspire to reteach the foundations of modeling. For that, a number of excellent, widely used, texts are available (e.g., for game theory: Austen-Smith and Banks 1999; 2005; Binmore 2007a, 2007b; Dixit, Skeath, and Reiley 2009; Gehlbach 2013; Gintis 2009;

Humphreys 2016; McCarty and Meirowitz 2007; Morrow 1994; Niou and Ordeshook 2015; Ordeshook 1986; Osborne 2004; Rasmusen 2006; Williams 2012). A separate set of books supplements the texts by demonstrating the usefulness of game theory and familiarizing readers with its intuitive aspects (e.g., Brams 2014; Dixit and Nalebuff 1993; Gates and Humes 1997; Shepsle and Bonchek 1997). As a hands-on, how-to guide, Fink, Gates, and Humes (1998) thoroughly cover a single theoretical topic. Morton (1999) addresses the role of formal models in the EITM program as a rigorous source of hypotheses for empirical testing; along the way, she provides many valuable insights into types of models and strategies for their empirical evaluation. Texts on game theory, social choice theory, and microeconomics offer students definitions, formulae, propositions, and exercises. For all of their strengths, such relevant extant works leave open a void: how to approach the real world with all these tools. As a result, many political and social scientists who wish to use formal models in their substantive research need additional resources that might aid them in that effort.

This book integrates formal modeling into the epistemology of the social sciences, just as it includes the tools of formal theory into research design, as a methodology. In so doing, it engages readers in a series of design demonstrations, ensuring that those designs reach into a variety of substantive research areas. Throughout, the book uses real-world stories to inspire the construction of formal models. It also levels the playing field, so to speak, for inevitable differences among readers by, for example, defining key terms upon first use. With this book, we address what we see as the need in political science and the broader social scientific community to better understand the place of models in research design.

1.5. Map of the Book

The next two chapters are devoted to the discussion of the epistemology and methodology of modeling. Chapter 2, dedicated to epistemology, elucidates the place of formal models in political and social science research. Chapter 3 delves into the methods and techniques of formal modeling. Along the way, it distinguishes the types of models that we design in this book.

The abstract theoretical presentation in chapters 2 and 3 leads to four story-based chapters showcasing how to put formal models to work. Chapter 4 illustrates the use of basic utility maximization, taking up a news story of traditional leaders in the South African countryside. Chapter 5 adopts decision theory to model regulation of the fishing industry by subnational and national agencies. In chapter 6, a single narrative gives rise to two types of models, social choice theory and cooperative game theory, capturing the strategic intent on the part of political elites as they negotiate an environmental treaty. Chapter 7 uses noncooperative game theory to address a story of a different kind of dilemma experienced by prisoners: abuse of power in the US criminal justice systems.

Alongside the differences in method and substance, the story-based chapters share a common structure. Each yields multiple narratives from a single news story, and then proceeds to formalize each narrative, refining the resulting models so as to reach a statement of ideas that is as general as possible. The distinctions among the models within the story-based chapters instantiate a key message of the book: just which sort of model will emerge depends on the question pursued by the researcher. The fundamental similarity in the process by which these chapters move from real-world events to nonformal arguments and then to models will appeal to scholars who want to incorporate modeling in their work.

The last chapter positions our work on designing models in chapters 4 through 7 as both the material evidence in support of, and the application of, our epistemology and characterization of the methodology of models from earlier chapters. In doing so, it summarizes the contributions of the book. We once again stress the essential capacity of formal models to test and build theory, and underscore the place of formal models in research design and their role in meeting the standards of analytical transparency, theoretical replication, and knowledge accumulation.

TWO

Epistemology

*A Path to Knowledge from Story
through Narrative to Model*

What is the epistemological place of formal models in social research? We have sketched how the analyst journeys from a real-world story to a narrative and then to a formal model. We have emphasized that a single story can inspire multiple narratives. Likewise, one narrative can provide the elements for multiple models. We have also stressed that we are working with individual actors and thus with micro-level theory and models.

Here we delve into the progression from evidence on human behavior (i.e., stories and narratives) to models, along with the overarching relationship between theories and all of these. These elements serve as the weight-bearing supports for the overall research design in any project we might pursue with the help of modeling. They offer the means of unifying into an integrated whole the key inputs in the development of formal reasoning. To the classical scheme of "conjectures and their refutations," then, we add that formal modeling establishes crucial linkages that reveal how to reach better conjectures and how to subject predictions to stricter criticism while drawing on both theories and evidence. These are the epistemological benefits of the use of formal (logical, or axiomatic) models in political and social science.

2.1. Formal Models in Research Design: A Closer Look

Figure 2.1 depicts the place that formal models occupy in research design. The entirety of this chapter details the relationships, as represented by arrows, in that figure. In figure 2.1, M denotes a formal model, T is theory, N is narrative, and E is evidence. Some element, E_i, in the set of Evidence is equivalent to a story and is a subset of E. As the figure indicates, the relationship between the theory and models is two-directional. Models (again, M in the figure) are designed with guidance from theory.[1] Yet, we contend, the process of design can (and should, as it is a form of critical assessment) later lead us to amend the theoretical argument. As shown in section 2.2, these amendments tend to change the initial theoretical argument so that it acquires greater specificity, that is, loses generality.

We build the model, M, out of the narrative, N, as shown by the horizontal arrow in figure 2.1. In turn, the narrative is constructed on the basis of a story (and thus from evidence), guided by theory. The narrative uses the moving parts of the story, that is, a single datum from the empirical world, E_i, in such a way as to conform with the theoretical argument. After the narrative does that work of synthesizing theory and observation, we formalize those parts in the model, M.

With E, the entire universe of evidence, models are also connected in two ways, although the nature of the model's relationship with evidence differs from its relationship with theory. As for the model's relationship with theory, the theoretical foundations as a rule should remain unchanged during the design process, even though modeling needs may push the theoretical argument to a greater level of specificity. The model links with evidence at two different levels: first, some single piece (or small set of pieces) of evidence "feeds" the model's design, and then other—multiple—data points serve to evaluate the model's predictions. Narratives permit processing of those initial bits of evidence prior to modeling. A narrative refines observation (data) by filtering it through the prism of the original theory, and in so doing leads the analyst toward model design. While the narrative-model block plays an epistemological role, the transition within it—from the narrative to the model—is more methodologically than theoretically guided.

Figure 2.1 distills how narratives and formal modeling enter into classical epistemology in political and social science. As far as we know, nothing similar appears in any published or unpublished work. Yet longstanding practitioners of formal modeling might recognize instantly that the figure conveys what they do and how they think. Indeed, figure 2.1 expresses in a

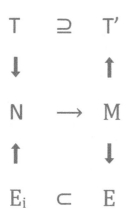

Figure 2.1. The place of formal models in research design: Multiple linkages among theory, narrative, model, story, and evidence

new way, for a new context, established insights from classical epistemology, while making explicit the story-narrative-model sequence, as noted.

Observe too that the figure incorporates an idea pushed throughout this book: the recounting of a narrative fuses theoretical and empirical elements, and occurs *before* the model can follow suit. Moreover, figure 2.1 suggests one of the book's recurring themes, already introduced, that multiple narratives can emerge from a single body of evidence, represented in a story. In these ways and more, the figure supplies the connective tissue for much of the book. We invite readers to hold figure 2.1 in mind as they encounter each specific example of model design in chapters 4 through 7, and work through the linkages between and among theory, narrative, model, story, and evidence for that specific example.

The bulk of this chapter is organized around figure 2.1. Yet we issue a caveat: the prose that follows suggests linear thinking, whereas figure 2.1 emphasizes the non-linearity of the journey that analysts make as they travel between and among theory, model, evidence, narrative, and stories. Caveat in hand, in section 2.2 we follow the arrow from theory to model (T → M) and from models to amended theory (M → T'). Section 2.3 connects models and evidence, tracing the links both from stories to models (E_i → M) and from models to evidence (M → E). Section 2.4 elucidates the role of the narrative, as the narrative, inspired by the story (E_i → N), builds a bridge from theory to models (T → N → M).

2.2. Why and When the Use of Formal Models
Improves Theories

Why would a researcher want to interject a model between theory and evidence in research design, as in figure 2.1? What makes having a formal model a good idea, and what sort of model might qualify as good? When might an analyst do without a model? These questions culminate in yet another. What are the advantages of a formal model, and under what conditions is one likely to be especially helpful to a student of human behavior as she pursues her work? This section addresses how a theory informs a model, and shows how formal modeling can improve theories. Here we elucidate what we previewed in chapter 1: modeling contributes to *theory building*, advancing the development of knowledge even before—or even without—the stage of testing hypotheses.

2.2.1. How the Theory Informs the Model: T → M

What is the relationship between combinations of the specific moving parts named in a theory, on the one hand, and the outcomes anticipated in the theory, on the other? How does posing and answering that question illuminate the linkage between theory and formal models? As noted, it has now become a commonplace that formal models can be hypothesis-building tools. If forced to separate strictly theory and evidence, most scholars would lump formal models with theory. Even so, they see models as "living" somewhere between theories and empirical analysis. Yet as the first chapter has suggested, models have a larger, independent role in the research process: they help build and refine theoretical arguments. Models help identify the outcomes that the theory, if all of its parts were engaged and working in concert, should produce.

As a social scientist commits herself to a theory, this theory dictates her bias in assigning relative importance to items of information contained in the raw description of events as well as in selecting a story in the first place as the basis for her narrative. This initial bias conditioning a scholar's thinking originates in her assumptions, which in turn are based in her theory.

A *theory* is a set of assumptions (statements) from which, once those are in place, a number of propositions logically follow. This definition is maximally inclusive. Occasionally, some scholars accuse others of using the word "theory" too loosely, as a label affixed to what otherwise could be called "unsubstantiated claims." Yet consider for the purposes of evaluating this definition that someone who makes unsubstantiated claims is simply

laying out assumptions about the way the world works. The latter is a legitimate course of action for a pure theorist. This definition of a theory still
applies when empirical analysis predominates. There, the term "theory"
refers to a mix of premises giving rise to a list of hypotheses that are then
evaluated against the evidence. Scholars at a minimum state their premises,
which might complement, or might compete with, each other. Even when
the assumptions underpinning the premises are left implicit, they exist and
they can, with some effort, be identified and articulated.

We acknowledge that some readers might not agree with us: How can a
theory boil down to a set of assumptions and logical extensions of assumptions? Such a question would reflect extensive exposure to theoretical arguments grounded in a combination of prose reasoning drawing on insights
from prior research and also prose logic seeking to move beyond prior
research. But, we argue, that modus operandi involves making implicit or
explicit assumptions and reasoning step by step on that basis.

Where does that leave the notion of a model, which is, similarly, a statement of assumptions and their logical extensions? The model holds constant the conclusion, as much as possible, and leaves open the possibility
that the premises may require a revision. The model differs from a theory,
then, in that it is a dynamic search for logical agreement between predictions and their premises, effects and their causes. During its construction,
the model is a flexible representation of a theory's intent. When successfully completed, the model becomes a replicable statement of the ceteris
paribus theory.

The model works to fill a possible void left open in a theory to the
extent that (some of) the assumptions are unarticulated, or few extensions
are derived, or both. Ideally, of course, one hopes that the theorist is clear
and aboveboard on both counts. In such a case, the model will simply replicate the theory without adding anything new. But often the extent to which
we can reason from a set of general assumptions is limited; sometimes we
can reach only a few nonspecific claims. Faced with such an impasse, scholars might push through to obtain operationalizable, falsifiable claims by
adding ad hoc speculations and restrictions. Yet if they do so, they depart
from the specificity embodied in their initial premises and obscure the path
by which their conclusions were obtained. When a formal model bolsters a
theory, the path from the theory to conclusions is not murky but rather is
clear, however mathematically cumbersome.

To appreciate the latter point, try thinking of models as answering the
question: If my stated theory were true, what would that logically imply?
This is no different from how we ask the question when drawing empiri-

cally testable propositions. In contrast to empirical models, however, when seeking the logical implications of a theory with a formal model, we turn not "outward," to factual evidence in the real world, but "inward," to the abstract component parts that together constitute our theory.

First, we identify what those are: assumptions. Then, we put them together and in our heads we force our actors as we conceive them to adhere to those assumptions, every step of the way. In other words, we build a model by "playing through" the process we have theorized, all along sticking faithfully to the theory-defining assumptions, making our actors march to the beat of our assumptions and seeing where that path takes us. Along the way, we may have to add further specificity—that is, more assumptions—and here is where the initial datum of a story serves to make the narrative and the model more detailed than the spare theoretical statement. The added details come from (some limited) observation, and so critics may question their general validity. Yet recall Gibbard and Varian (1978, 665): "Assumptions cannot meaningfully be called 'unrealistic' without more said, for the same model can be applied to many different situations in the world, and its assumptions may be realistic for some of these situations and not for others." Even if a model's assumptions are (perceived by someone as) unrealistic, they are still of use for understanding the world. Hence, by adopting the story-model sequence, we ground assumptions for our model in the story, and thus achieve (reasonable) empirical accuracy for that particular observation at least. If in the course of building a model we discover the need for some assumption that is not elucidated in our story, that means that we must reach out to other stories or to prior models, or conduct an analysis while allowing for each possible version of a particular assumption. Analysts make assumptions not because we *know* what the right assumptions are, but because otherwise no analysis would be possible. In formal work, we make assumptions explicit, and we are open about the fact that a theory or a model is just a creature of the assumptions made.

This is the road symbolized in figure 2.1 as T → N → M. For now, we abbreviate that to T → M, because the interposed N (narrative) is itself linked to additional inputs that extend beyond the theory. We deal with those inputs later.

2.2.2. How Models Test and Build Theories (M → T′, Part 1)

Playing through the process within the construct of a model generates the invaluable output of (combinations of) strategies that actors *should* adopt in equilibrium (or in multiple equilibria).[2] This type of output from the model is reflected in the observed (in the model, not yet in reality) behavior of the stylized actors. It may correspond to the real-world behavior of actual political actors as well. The second type of output from a formal model is the outcome or outcomes that it predicts will take place according to the solution concept that the modeler employs. These outcomes occur deterministically in the universe that is constructed from, and guided by, the "laws" of that particular theory and its constituent assumptions; this determinism applies even if we predict precise probability distributions over specific contingencies. It just might be also the case that corresponding substantive outcomes actually occur in the universe of observed reality.

At this juncture, a modeler has something to compare, since she has established some parallels across two domains, the modeled one of logical steps and the theoretical one of premises and conjectures. She can now pose questions about degrees of similarity and difference between corresponding objects in those two places. *Testing* and *building* theories takes place in the course of such a comparison. This we consider a form of critical assessment, thus a test. Consider that in the expert analysis of Hands (1991, 114), even Karl Popper's critical realism contends that "the 'method of science' consists of criticism and being open to criticism. . . . On this view, empirical falsifiability is simply one of many forms of criticism."

Figure 2.2 represents in a stylized way the possibilities that may arise from this process of comparison. It lays out the contingencies on the path that a designer of models might travel as she constructs a model to both test and build theory, and thus probes T → M → T′, as in figure 2.1. In what follows, we track, with readers, each step of the way in this process of comparison between the modeled domain of logical steps, on the one hand, and the theoretical domain of premises and conjectures, on the other. Throughout, our purpose is to clarify how a formal model serves to *test theory*, and how, when tests return unexpected results, additional investigation of the model *builds theory* as well. We work through—walk through—figure 2.2 as it outlines our proposed algorithm for evaluating (critiquing), and adjusting, the mutual fit between these two abstract domains.

As the multiple quadrants of figure 2.2 indicate, the initial formal predictions and the anticipated theoretical argument may or may not resemble each other. First, the formal model may have already dealt our theory what might seem to be the ultimate blow—if the model has failed to generate any solution, that is to say, the solution does not exist. If in our model we stuck faithfully to our theory's premises, then this nonexistence means that those premises do not logically lead to the anticipated conclusion—in

Type E test / Type L test	Meets	Fails
Meets	I. Translate model's outputs into hypotheses	III. Return to the model to reframe the question: Under what additional assumptions would an equilibrium exist?
Fails	II. Return to the model and ask the question: Do there exist parameter values for which the model's predictions equal theoretical predictions?	IV. Choose a different research question

Key:
E: Existence of a solution
L: Logicality

Figure 2.2. Models fulfilling their roles: Theory-testing and theory-building

fact, to any conclusion at all. A situation of this sort often arises when an argument first meets a model. Even in multiple iterations, with subsequent ones becoming more developed, many modelers find that attaining the intended solution concept just may not be possible. The worry now is that the theory-based modeling world may yield no prediction. To discover this failure via a model would mean that the theory in its current form has also failed. Here we grasp a model's role as a test of a theoretical argument. We consider passing this first hurdle a *test* (of a theory by the model) and call it the test of *E*xistence of a solution (labeled *E* in the figure).

What if the Type *E* Test (Existence of a solution) is not met? Should the scholar give up when she learns via her first sketch of a model that *E*xistence fails, or should she adjust assumptions to make things work? Observe: this is not an empirical test of hypotheses that we have failed; instead, we have failed a comparison between two cognitive constructs. If our first attempt at a model has failed our theory on *E*xistence, this must compel us to hunt for the conditions, parameter restrictions, and modifications of assumptions that would ensure that some prediction emerges in a model. How distant the resultant construct would turn out to be from the original theoretical premises is a matter of luck and circumstance. If, for example, our original theory was too general and so embraced cases that were too diverse to support any consistent prediction, then we can fix the problem by narrowing the set of cases. This does not mean that the theory is corrupted or that we have crossed the line into data mining and dropped our original theory altogether. Rather, we have discovered that our theoretical argument is not as broadly applicable as we initially envisioned: we have identified scope conditions. The theory has either remained unaltered or acquired greater specificity in the critical process as in figure 2.2.

It is thus possible that a model failing the Type *E* Test may still be "fixable" after its initial scope has been narrowed (and so *Existence* has been reestablished). To preview the discussion in section 2.2.3, figure 2.3 below identifies strategies for "fixing" a game-theoretic model so its next iteration meets the *Existence* criterion. If we manage to correct this problem, and *Existence* is met, then we would find ourselves in the top row of figure 2.2, and continue along that newly acquired contingency.

Observe that a model may "fail" a theory in a second way: the model succeeds in making a prediction, but its prediction differs from what the theory holds should happen (or has happened if the theory engages in postdiction, as often applies in political and social science). What if a theory were to hold that agents should successfully work together, for instance, but a model for that theory were to return a deterministic prediction to the contrary—of inefficient, mutually undermining behavior? Such is the "everybody gains" argument for why people would all contribute to produce a public good; this theory is fully rejected within the model of collective action based on individual rationality. Even though the *Existence* test is met there, the equilibrium prediction is free-riding. When we focus on a model's purpose in exposing the logical fit or flaws of theory (as in the scenario just seen), we submit theory to a Type *L* Test (*Logicality*).

Thus, having established via a model that a solution exists, that is, the *E* test is met, we then must move on to compare the model's initial formal predictions with its starting theoretical expectations. Depending on the outcome, we will arrive either in Quadrant 1 (Q1) or Q2. This comparison between what we think our theory implies and the received output from a formal model built with our theory's premises resembles the use of simulations in empirical research. In the case of formal modeling, though, our simulated space is the world within the model and the logically feasible behavior in that world. Just as with statistical simulations, behavior implied by the model serves as faux data that either will or will not conform to prior theoretical expectations. The former contingency corresponds to Q1 in figure 2.2, the latter to Q2, both dependent on having met the *Existence* criterion as a necessary condition.

When our model has passed the *E* test, and yet it has failed the test of *Logicality*, we land in Q2. In such a contingency, we can take steps to reconcile the predictions. These steps are further elucidated in section 2.2.3,

with figure 2.3, below. For now, we emphasize that this work of reconciliation clarifies which assumptions underpin the desired results and so
clarifies as well how to qualify, that is, restrict, the theoretical argument.
While testing for *Logicality*, we are not confronting either model or theory with the empirical universe. Instead, we are improving conjectures
for future empirical assessment by figuring out with greater precision
what our theory does and does not imply. Thus improved, conjectures
become theoretically replicable. To clarify: it is now assured that any
scholar can reach the same conjectures from the theory's stated premises.[3]
We disagree with those scholars who understand theoretical replication
as merely a subtype of empirical replication. We fully embrace, however,
the insistence on transparency and clarity in technical exposition (e.g.,
Lupia and Elman 2014).

If, despite all attempts to adjust the model's assumptions, the formalization fails both the *Existence* and *Logicality* tests, we reach Q4 in figure 2.2: it is the end of the line for this research question. The option
presented in Q4 for the researcher is to choose a different question to
explore, because reframing the current one has failed to yield results.
Otherwise, our efforts would have eventually brought us via the iterative path to the gratifying Q1, where we have a theory (T′) that meets
both the criteria of *Existence* and *Logicality* in successfully predicting the
outcome. In such a sweet spot, in Q1, we may then turn to the empirical
universe and see whether our model generates hypotheses that can be
subjected to empirical testing.[4]

This is how models "test" theories—how they improve the clarity of the
path to conclusions from assumptions and the range for which the conclusions hold. Models either work or not, and if the theory gets amended so
that the model works, it becomes a better theory. Theories that are just
stated assumptions (e.g., "people are rational") are not replicable. Theories that are assumptions along with presumed conclusions ("people are
rational, thus markets are efficient") are replicable to the extent to which
they are modeled. If the "test by modeling" has failed, the theory has failed
to replicate. To have a replicable theoretical argument in our view means
to have developed a mechanism—a deterministic model—in which the
logical bridge from assumptions to implications as in the original theory
is obtain*able*. As soon as the analyst manages to construct at least one such
mechanism, the theoretical argument becomes replicable. To illustrate,
"people are rational, thus markets are efficient" is replicated via the general
equilibrium pricing model.

2.2.3. Strategies for Reconciling Theory and Model (M → T′, Part 2)

To reiterate, if the model fails to align with the initial theory, formal reasoning gives a systematic way for structuring thinking and critically refining theory step by step, rather than dismissing (rejecting) it outright. Because this stage comes before empirical testing, using models to amend theories does not involve "cooking the books" where the book is the body of evidence. Consider an experienced engineer who in the face of a real-world severe structural defect requiring an ingenious solution must grapple with both the fundamental principles of engineering and the inferior construction of the machine. The principles she has internalized as an engineer must guide her at every moment. Yet the possibility of finding a solution may hinge on expanding the set of characteristics, that is, incorporating additional parametric bounds, which are not part of the standard design. The engineer would adjust one variable at a time until the solution would overcome the structural defect. The test of the engineer's amended theory is the machine that is fixed according to that theory—the prediction that works (cf. Putnam 1991).

The question that we use to evaluate our initial theory differs across Quadrants 3 and 2 in figure 2.3. In Quadrant 3, we seek an answer to "Under what conditions does some prediction hold?" In Quadrant 2 we ask, "Under what conditions does the 'right' prediction hold?" We now detail the advice summarized in bullet points in figure 2.3 for what to do when the *Existence* and *Logicality* conditions are not met. Having identified the contingencies represented in Quadrants 2 and 3 in figures 2.2 and 2.3, we move to "fixing" the problems that placed us in those quadrants, and adjust the fit between theory and model, hopeful of reaching Quadrant 1, where the two work in harmony.

As seen in figure 2.3, all of the advice boils down to adding assumptions and restrictions to the model. Adding even one more assumption always reduces the generality of an argument. The argument now applies only if some additional conditions are met. When we add assumptions, we add constraints, that is, parameter bounds, and so make the argument into a conditional one and the theory's prediction into a qualified one. To heed Morton (1999, 65), "the additional specificity [of the assumptions] will make solution of the model easier but needs to be considered carefully."

Nonetheless, engaging in this exercise offers many lessons. It permits us to say when the argument does apply as well as when it *does not* apply. The new knowledge yielded can be of interest and of use in many contexts. It also points the way to new propositions, some of which could be empirically testable.

Type E test / Type L test	Meets	Fails
Meets	I. Translate model's outputs into hypotheses	III. Steps to take* • Mixed Strategy NE • Continuous strategies • Additional moves • Additional players • Information
Fails	II. Steps to take • Treat the model as an equation • Derive domains where *L* is met • Specify predictions outside of the resulting domain(s)	IV. Choose a different research question

Key:
E: Existence of a solution   MSNE: Mixed strategy Nash equilibrium
L: Logicality   NE: Nash equilibrium
* Specific steps depend on the methodology of a model; the list in Q3 illustrates the steps for non cooperative games.

Figure 2.3. Iterative reconciliation of model and theory

If we add such new assumptions one at a time, then each addition constitutes a necessary condition for the theory, that in its previous state, T,
has failed the E Test to become the amended theory, T′, which passes the
E test and "works." The method consists in treating the model as an equation, where the original, insufficient set of assumptions gives the known
values. Another known value that we can "plug in" is the theory's original
prediction—the one that we have failed so far to recover via formal modeling. On the other hand, the unknown values, which should balance the
equation, consist of a missing parameter or set of parameters, which, once
derived, will become the additional assumptions.

If *Existence* of a solution is in question, then the model thus far
has failed to make predictions of any sort. Depending on the choice of
methodology—decision-theoretic, noncooperative game-theoretic, or
cooperative game-theoretic—the corresponding solution concepts have
different levels of robustness. Chapter 3 discusses solution concepts and
thus modeling methodologies. Observe here, in accordance with Quadrant
3 (Q3) of figure 2.3, that the most immediate fix to establish the *Existence*
of a solution is by *adopting a less restrictive solution concept*. While across the
variety of formal methodologies surveyed in the next chapter the solution
concepts differ *in principle*, within each method there also exist variations
on the main approach, some "harder" to attain, and others, "easier." One

example is the difference in the difficulty of attaining *Existence* in a model of a Pareto-efficient[5] Nash equilibrium, versus "settling" for establishing the *Existence* of a Nash equilibrium (i.e., without any additional requirements to it).

Looking more closely at what to do if we find ourselves in Q3 while solving a noncooperative game, one of the bullets there suggests that the scholar allow *Mixed Strategy Nash equilibria*,[6] whereas initially perhaps she was looking for a Nash equilibrium in pure strategies. "Mixing" strategies when solving noncooperative games generates predictions that are probabilistic, and is equivalent to creating continuous strategy spaces from an initial finite set of strategies.[7]

Three clarifications are in order, on the existence of equilibrium, the nature of empirical expectations, and the meaning of continuous strategy spaces. First, as a general theoretical result, the Nash equilibrium always exists in continuous strategy games. Hence, to the extent that a scholar can justify her actors using lotteries (probability distributions) over their pure strategies, thus moving to adopt mixed strategies, she can always achieve equilibrium existence.

Second, however, this sort of achievement comes with a caveat: when the predicted solution is an equilibrium in mixed strategies, the empirical expectations from such a prediction can be extremely obscure, for example, can stipulate that actors might do different things, unpredictably, in any single play. The outcomes as produced by the combinations of such unpredictable actions would thus vary across the broadest range. Unless the play is repeated a substantial number of times and we are thus able to observe the resulting frequencies, the prediction of a particular mixed strategy equilibrium would be nonfalsifiable. Some weaker form of verification might be available (e.g., Carnap 1966).

Third, designing the game to give players *continuous strategy spaces* technically resembles allowing for mixed strategy equilibria as a solution concept. This, however, has the advantage of avoiding a nonfalsifiable prediction while ensuring that a Nash equilibrium exists. As an example of a continuous set of choices, consider solving for the quantity of output of a particular good to put on the market, while knowing the demand among consumers, in the form of a continuous demand function. That would generate a prediction (e.g., some number of tons), and we would very soon discover whether it was correct (in corresponding to actual consumer demand) or not.

Further proceeding through the steps in Q3 of figure 2.3, allowing players to have *additional moves* increases the complexity of the model, which may make it less tractable. Yet this complexity may reflect the assumptions that were at first omitted but turned out to be essential to bring strategic actors to behave in certain ways. Adding moves alters strategic calculations, and so affects the findings with regard to both *Existence* and *Logicality*—both whether there is a predicted solution and what it entails.

Introducing *additional players* produces a major complication for analysis. But their presence in a model might push other players to choose specific behaviors and that would drive the model's solution or solutions. As with adding moves, adding players affects both the *Existence* and *Logicality* of the model's predictions, and therefore this suggestion can also be used for the circumstances in Q2.

Information in the model is a very important parameter, and the scholar can manipulate this parameter and achieve dramatic reversals of solutions and predictions, as desired. Models can serve to uncover that certain predictions are contingent on very specific assumptions about information. Chapter 3 discusses this further and chapter 7 shows the use of information in designing models.

When we restrict parameters and thus make our model less general, corresponding changes can fall into several broad categories. For example, insofar as noncooperative games are concerned, the bulleted list inside Q3 gives a modeler the steps by which she can add new parameters or new restrictions on included variables, can change the information and risk structure, incorporate unobserved costs to players, account for long-run expected payoffs, and even expand the game itself to encompass more choices, more actors, and more contingencies.

When does the researcher stop bringing in more variables and give up on attempts to "fix" the theory? This is a vexing question, and if the method were exploited to the extreme, then we would have to concede that Green and Shapiro (1994) have a point. To Green and Shapiro (1994), Hands (1991), and others, the individual rationality principle, which is the shared foundational assumption in economic models, threatens the Popperian falsification program in the social sciences. The process of adjusting models can continue ad infinitum, which raises the specter of "overfitting." Yet empirical statistical modeling has the same problem. Scholars occasionally work *too* hard to fit evidence to their hypotheses. Data mining—adding arbitrary control variables and interaction terms in order to improve the fit—is widely acknowledged as the culprit there. Overmanipulating the parameters of a formal model in some ways parallels overmanipulating the parameters of a statistical model. Likewise, just as formal models may be tweaked to produce alternative predictions, so statistical models can yield diametrically opposite predictions depending on assumptions and specifications (see, e.g., Cingranelli and Filippov 2018a, 2018b).

We will be satisfied that we have arrived at a sufficient set of assumptions (or constraints) when we have managed to land in Quadrant 1, meeting both the Type E and Type L Tests. As soon as we have narrowed our theory to ensure that a prediction can be made (to satisfy *Existence*), we must move to the scenario where the Type L test fails, and we reconcile theory and model so as to satisfy *Logicality*. The discussion proceeds, then, to the bullets in Q2 in figure 2.3.

The bulleted list in Q2 spells out possible steps to rectify the failure of the type L test. In brief, in addition to considering the amendments to assumptions already discussed for Q3, we should now seek to derive the parameter domains where the insights from the theory logically hold, and we do so by *treating the model as an equation*. To treat a model as an equation means to presume that the prediction is known, but some of the underlying parameter values (assumptions) are not. The key is to suppose that the original theoretical prediction is met, and to parameterize the moving parts in the model, looking for parameter values where the "equation" holds, that is, the expected prediction is sustained. A model may undergo revision in the course of this exercise with regard to the included actors and actions, information restrictions, as well as restrictions on the domains of parameters that describe payoffs and probabilities.

Once *Logicality* is reestablished, even if only for some limited parameter values, the scholar should also specify the resulting model's predictions outside the identified domain(s). With that done, she is not only able to say when she expects the argument to apply, but also what should happen when the specified conditions are not met. Between the two statements, there may emerge substantial opportunities for empirical testing. To emphasize: if and when a researcher successfully aligns model and theory, she either has adjusted the model and managed to reproduce the theory's initial prediction(s), or has adjusted the theory to amend its predictions by deriving the corrected logical implication(s) of the theory's original premises.

What if we do not succeed in restoring *Logicality*? In this case, no parametric bounds exist for which an initial prediction can be recovered. We have just discovered that our initial theoretical reasoning was and is thoroughly faulty. The conditions characterizing such a case mean that the attempts at the bulleted fixes in Quadrant 2 in figure 2.3 have failed, and it is the end of the line for that argument. We find ourselves then in Quadrant 4 of the figure, and need to start thinking of an alternative research question.

In sum, figure 2.3 presents the logistics of the *iterative* relationships between theory building and modeling, which correspond to M → T′ in figure 2.1. To present the synopsis: if the model meets the requirements of existence (*E*) as well as logicality of the original prediction (*L*), then we arrive in Quadrant 1 of figures 2.2 and 2.3, where we have successfully verified the theoretical argument. On the other hand, if either *L*ogicality alone or both *E*xistence and *L*ogicality are in question, then we do not have a valid argument. The process of reconciliation just described is not rapid or straightforward by any means. In fact, it may require us to move back and forth many times. We add or adapt assumptions until our theory meets the *E* and *L* tests.

At this point, it is fitting to invoke Hausman (2009), who characterizes "the fundamental theory of microeconomics as a set of inexact laws. . . . What makes these laws inexact is that they contain *ceteris paribus* clauses" (Hausman 2009, 41). Reformulating theory after modeling refines T and amends it to T′ with such ceteris paribus clauses that make the theoretical prediction(s) valid logically.

2.2.4. Limits of Reconciliation between a Theory and a Model

If a test were to fail while we were testing hypotheses by statistical means, we would reject the hypotheses. That would be the end of the road for those hypotheses and for the theories from which they stemmed. But we should not likewise reject a theory at the first instance when it fails a test of formalization. Again, the theory *building* capacity of formal modeling lies precisely here, in the process of reconciling theory and model. Be forewarned: it is just this capacity, perceived as injudiciously used, that has elicited the strongest critique of the formal method to date (Green and Shapiro 1994). We on the contrary choose to celebrate this power of modeling. Models provide a path to amending a theory *before* we subject it to any test against evidence. It is this path we indicate in figure 2.1 and retrace in figures 2.2 and then 2.3.

We are not cavalier about amending theory when it fails in its initial formalization, and we are quite aware that there is no magic formula for avoiding altogether the pitfalls of "overfitting" a theory (i.e., making too many assumptions) in the pursuit of attaining results. We thus offer the reader the algorithm by which she can proceed cautiously, step by step, to amend a theory to some limited and minimally necessary extent at each iteration of modeling. To amend a theory is to amend its assumptions.

Models are the method by which we can derive the missing assumptions and (or) render explicit what were once implicit assumptions.

For formal models, just as for statistical models, the issue of "assumption mining" (to parallel "data mining") would pose a valid methodological concern. Both leave us without a theoretical foundation on which to build further inquiry, even if both processes establish the predictive power of models. One difference between formal versus statistical "mining," however, is that the former can still be followed by a test of external validity, as it does not use up the empirical evidence in the process of model-fitting. Evidence is a finite resource, consumed in the process of statistical mining. Whatever fitting proceeds in the process of formal modeling does not exhaust the evidence: the "fitting" in a formal model is performed *before* any foray into the empirical universe. Thus, we argue in favor of taking the assumptions-fitting approach to theory-building as an appropriate form of theoretical work.

Even so, worrisome possibilities exist. Specifically, we flag as highly problematic any amendment of the core premises that define a theory as a consequence of fitting the model. Should such an amendment occur, the term "theory-building" no longer fits the enterprise. Something along the lines of theory-revisioning (not misspelled, for the term is based on "revisionism") comes to mind. For this reason, we would be instantly concerned when scholarly work jumps to a model directly, without committing beforehand to a theoretical family, if not a theoretical genus. The jump may indicate an atheoretical answer-fitting exercise. Although we are concerned with such a practice, we by no means intend to condemn all papers jumping straight to the model and bypassing theory! Rather, rooting the formal model in extant broader theory and introducing a finite set of well-articulated additional assumptions should be considered a good sign, especially if the scholar then points out as one of her findings that her work shows the limits of the broader theory's applicability.

2.3. The Connection between Models and Evidence ($E_i \rightarrow M$ and $M \rightarrow E$)

Turning to the connection between models and evidence, the first thing that strikes the reader is that the arrows in figure 2.1 go both ways, encompassing an element of induction ($E_i \rightarrow M$), as well as conventional hypothesis testing, which, when joined to modeling, entails deduction (the reverse, M → E). Models thus link to the empirical world twice, both taking information from the real world as inputs for their construction, and offering predictions that can be subjected to empirical testing. Given the use of observation in constructing narratives and models, the two-way path winds through an inductive stage, and yet this induction serves a deductive purpose: it increases the specificity of theoretical predictions and so the specificity of the resulting propositions.

2.3.1. Modeling Life, the Universe, and Everything (E_i → M)

Evidence and theory are both inputs for modeling. Political and social scientists use their knowledge of facts to obtain their models' initial structure. This is where they locate the basic moving parts of a model and the labels they attach to these moving parts. Specifically, the basic elements of a model include (a) actors, (b) their available moves, (c) the resulting outcomes, (d) the actors' preferences over such outcomes, and (e) the information that actors have about each of these, including the information available to all actors. Rubinstein (1991, 909, emphasis in original) argues that "a good model in game theory has to be realistic in the sense that it provides a model for the *perception* of real life social phenomena."

Observe that if a model were purely mathematical, as in, for example, game theory as a subfield of mathematics, there would be no actors at play. The mathematician needs no actors when she can directly specify those functional forms that mean actors' utility functions to a social scientist. In mathematical game theory, the game starts with the matrix of a strategic form, and derives the mathematical properties of the objects in the matrix. This suffices to explore equilibria as they are mathematically defined, without ever mentioning any actors or indeed anything of substance. Since any permutation within a class of such matrices constitutes a distinct game, a mathematical game theorist would also engage in characterizing the subclasses of such games that exhibit mathematical properties of interest. A mathematician might go on to explore the attributes shared among some subclasses of matrices—games. Had any of that been our purpose, then we would not need evidence to identify the game's components: we could simply specify all possible permutations as stated. Games so constructed could hardly be called models, however. Even though an analyst could draw analogies from them to the empirical world and borrow a purely mathematically devised game to serve as a model for some phenomenon or other, a mathematical game, in its *con-*

struction, is entirely self-contained and does not constitute an abstract representation of anything. It does not possess social meaning. We as social scientists must attach conceptual interpretations based on observation before we can use a mathematical model.

The mathematical model, so constructed, is akin to an arithmetical expression that appears on Tom Sawyer's math test, but that also in principle could be directed to a distinct purpose, for example to calculate the number of remaining apples in Aunt Sally's cellar. In the first instance, it constitutes pure theory, just as anything in mathematics does. In the second, if Tom were to come up with an answer from his own experience, it would be inductively derived. By itself, the expression is not a model. In conjunction with stealing apples, it is. It therefore synthesizes the induction from empirical evidence with the structure and symbols of mathematical theory. Note too that when Tom counts the leftover apples, he does not scan his last math test in search of a device to use as a counting model. Instead, Tom constructs a similar but separate device, a model by which he would know how many apples remain, considering that he pilfers them one at a time. The variables in his arithmetical expression will now have meaning and names, beyond their symbolic mathematical existence, and the expression itself will constitute a model.

Much like the fictional Tom Sawyer, in substantive political and social scientific research we care not about the class of mathematical objects but the analog of apples—the class of social "objects." In the study of human behavior, models are abstract representations of the objects we study, not themselves objects to be researched (cf., e.g., Rubinstein 1991). Thus, the model's parts must correspond to some perceived aspects of the object of inquiry (cf. Carnap 1936). In the epistemological schema in figure 2.1, this information about the model's moving parts comes from some initial observation of the phenomenon of interest—from some useful context in which we can pose our question or confront our puzzle. The manifestation of the phenomenon of interest is the first appearance of the empirics, the one that precedes the construction of a model.[8]

Here we use the term *stories* to denote those singular empirical manifestations serving as starting facts for building models. Stories are subsets of observed reality—of the world of human interaction, since that is our interest as political and social scientists. What makes a useful context for modeling out of a naturally occurring story, however, is the transformation of the story into a *narrative*.

2.3.2. What Is a Narrative and How Is It Constructed?
$E_i \rightarrow N$ or $T \rightarrow N$?

As chapter 1 previewed, a narrative transforms a story, a set of real-world events, into a piece of evidence relevant to a theoretical argument. Observe that figure 2.1 depicts the narrative as a creature of both evidence (story) and theory: $T \rightarrow N \leftarrow E_i$. The narrative draws out and strings together a subset of the facts in a story, departing from specificity by ignoring those facts that play no role in generating the outcome—given the narrator's premised theoretical foundation. In doing so, a narrative is already a model of the story, selecting theoretically opportune facts. Narratives stylize, to varying degrees, accounts of social phenomena that pose analytical questions. For a researcher intending to build a formal model, the narrative serves as the precursor to a model.

Our treatment of narrative broadly fits in a diverse array of understandings of and approaches to narrative in the social sciences. Thus, Hinchman and Hinchman (1997, xvi) define narratives as "discourses with a clear sequential order that connect events in a meaningful way for a definite audience and thus offer insights about the world and/or people's experiences of it" (cf. Elliot 1999, 2005; Wengraf 2000). In their survey of theories of narratives, Patterson and Monroe (1998, 319) underscore that the literature, though variegated, converges on the "critical role of narrative in the construction of meaning." The construction of meaning, in turn, is a necessary step in the acquisition of knowledge: as Carnap (1936, 420) stresses, establishing the meaning of a statement must precede both its confirmation and testing.

Our epistemological approach distills the idea of a narrative to a formal notion of narrative as a subjectively selected (biased) *subset* of a set of facts constituting an event. This definition is consistent with the role played by narratives in our epistemological schema. Thus defined, it is clear how a narrative processes the evidence for the purpose of formalizing a theoretical argument. This gives narratives their due as intermediaries between the world of facts and the realm of conjectures.

As we previously emphasized, the narrative is inevitably biased in favor of selecting those bits of evidence that conform to the narrator's implicit or explicit theory. Therefore, the narrative inevitably trims factual material irrelevant to the initial story. Any narrative limits the story it recounts to what the narrator sees as the relevant moving parts and the interactions among them, so that the narrator can either imply or directly proclaim a causal connection between actors' choices and the outcomes observed.[9]

In this we extend but also challenge Rubinstein's (1991, 919, emphases in original) view: "A game-theoretic model should include only those factors which are perceived *by the players* to be *relevant*. . . . This . . . makes the application of game theory more an art than a mechanical algorithm." For us what is "relevant" is driven by theory and thus is subjective to *the theorist*. The subjectivity of what is relevant explains the multiplicity of feasible narratives and models, all based on the same story. The "art" is when the analyst constructs the narrative, moving from E_i to N ($E_i \rightarrow N$). Designing the model from a well-constructed narrative, to us, in fact resembles a mechanical algorithm rather than art.

Within the narrative, a researcher can now construct relationships among the chosen facts on the basis of particular theoretical premises. This step already transforms the narrative into a theoretical model, whether formal or not. The narrative's elements supply the moving parts. That model frequently takes the shape of a causal explanation of some outcome (i.e., one of the narrative's elements). An explanation is composed of the facts privileged by the analyst (other elements of the narrative) and might be generalizable to a testable conjecture.

Yet various pitfalls might arise in such theoretical structures. When ad hoc assumptions creep into the narrative, another researcher studying the same phenomenon will diverge in the conclusions she reaches from her narrative. Adjudicating the clashing conclusions will become impossible. To avoid the pitfalls, the analyst must adhere to the same assumptions throughout. The use of formal models guarantees the scholar control over, and transparency in, the axiomatic structure.

Whereas a narrative is a biased representation of an event—a theory-tailored "just-so" story (recalling Elster's [2000, 693] characterization of "much of applied rational choice theory")—it leads to a model that makes its bias explicit and transparent. That is, relevant elements in a narrative give rise to the stylized components of a formal model. And, as argued, the model contains more than was supplied by either the theory or the set of events, because it fills in omissions in the original theory's logical argument. Hence, as argued, the model both tests and extends the original theory. The model also supplements the set of available facts, exposing the unobservable or unobserved aspects of the story.

The model advances the work of the narrative in another way as well, by enabling generalization. Through induction from a single story to an abstract form, the model puts the observed events on a continuum of similar yet distinct other possible events that fall in the same general category, a category itself characterized by the model. This is achieved when the modeler first makes explicit the assumptions about the rules of interaction among actors, and then analytically reveals the bounds of the parameters within which a particular type of dynamics (e.g., certain equilibria) would unfold within her specified structure. It is given those rules and within those parameter bounds that the prediction is confined. Within those rules and outside the parameter bounds, a prediction is *known* from the model *not* to hold. We do not and cannot know the entire remaining unmodeled universe, and make no prediction for it. The model thus makes explicit the limits of a given theoretical argument's applicability by amending the set of assumptions beyond the original theoretical premises. The model also limits the universe of empirical observation to which its set of theoretical predictions applies.

2.3.3. Thick vs. Thin Narratives: The Art of Finding the Balance

How broad or narrow a subset of facts should a narrative be? A narrator does not need to discard each and every aspect of a story that does not contribute to illustrating the theoretical argument she is developing. Few narratives are as spare as they can be, and some extraneous facts are more likely than not to remain. This excess factual material becomes valuable when a formal model's structure calls for adjustment because something in it does not work quite as initially expected. Indeed, since there might be times when we would need to increase the specificity of the model, as in figures 2.2 and 2.3, we would prefer a thicker narrative to a thinner one. Were we to omit everything that seemed not theoretically pertinent, the modeler would have no room for expanding the model and experimenting with alternatives. Put differently, if we "knew everything," then the thinnest narrative would itself suffice in lieu of a model since a model based on it would be guaranteed to work at first try. To clarify further, formalization would not generate any additional insights. The point, however, is that we do not know everything. Proceeding to a model is meant to test our theoretical argument.

On the other hand, mimicking life too closely, incorporating various bits of observation into the model's formal structure, would lead to over-specification. The more we "account" for the intricacies of the operation of the real world (more precisely, of a story from the real world), the fewer other stories can possibly conform to the resulting formal structure. Over-specification of a formal model limits generalization. Excess detail reduces the number of suitable data points (facts) that might be fruitfully explained by the model and serve as additional data to which the argument applies.

At the later stage represented in figure 2.1 as M → E, overspecifying the model would weaken our search for external validity.

The cost of overspecification is the same in formal and statistical models: vanishing degrees of freedom. In other words, we might have added too many assumptions, assumed too many specifics in order to obtain the desired "prediction," and as a result have in hand a model that applies to one unique case and no others. Put differently, the dynamics in the model are contingent on so many and such restrictive assumptions, that they are simply not met anywhere beyond the single starting case. When other applicable cases are unlikely to exist because the assumptions are too tailored to the concrete set of circumstances used to construct the model, there is nowhere for us to probe the external validity of the model's predictive strength. Our so-called explanation then would approach a tautology: the outcome in the narrative is "like that" because the circumstances in the story—and thus the narrative—are "like that." The model then would succumb to the criticism encapsulated in the label of a "just-so" story (Elster 2000, 693).

2.3.4. Overextending the Inductive Step: The Special Case of Fusing Narratives and Models in the Method of Analytic Narratives

Our epistemological view of narratives and models differs from the well-known, highly influential analytic narratives approach (e.g., the pioneers, Bates et al. 1998, 2000a, 2000b; Schofield 1999, 2002; cf., within and beyond political science, e.g., Carpenter 2000; Crettez and Deloche 2018; DeLong 2003; Nalepa 2010; Pedriana 2005; Schofield 2006; Zagare 2009). The key distinction regards the length of the inductive step. Most social scientists have read and enjoyed essays that analyze a historical case of particular interest and parameterize it in a model. Such essays capture everything of presumed significance for the final outcome in the model's structure. The resulting model is uniquely tailored to the case in point, and, because of that tailoring, it does justice to the complexity of the strategic path followed to the outcome by the story's principals. Practitioners of this methodology should be invulnerable to the criticism that "the real world is not like that!" Yet even they cannot satisfy everyone: both their narratives and their models inevitably prioritize some aspects of reality over others, due to their theoretical biases (cf. Elster 2000).

The prominent methodology of analytic narratives is a special and extreme case of the more general relationship that we describe between narratives and models. Our conceptual structure admits a view of the analytic narratives approach as a variation on the schema in figure 2.1. Analytic narratives move E_i → M, but not M → E: they do not engage with the generalization stage. In the analytic narratives approach, narratives and models flow seamlessly into each other, epitomizing abductive reasoning.[10] Unlike Bates and colleagues (1998, 2000a, 2000b), we exploit the possibility and process of revising theory with the aid of modeling. For Bates and team (1998, 16), it is "the construction of analytic narratives [that] is an iterative process," one they see as akin to process tracing in Alexander George's formulation (George and Bennett 2005). As we harness narrative to both epistemological and methodological purposes, we do not see or treat the construction of narratives as iterative. Rather, the construction of a model from the narrative is iterative; and the refinement of theory with the aid of modeling is iterative as well.

We seek narratives that are *multiple* renditions of the same story, leading to different models depending on the underlying questions of interest. In contrast, analytic narratives "seek to piece together the story that accounts for *the* outcome of interest" (Bates et al. 1998, 11, emphasis added). An analytic narrative is the single best way to represent the puzzle and the dynamics in the story. In this approach, details that are idiosyncratic to the event in question serve as pivots to the strategic action of players. Explanation often stresses the uniqueness of encountered parameter values and the importance of thorough knowledge of context.

We want to recover the outcome of interest, just as Bates et al. (1998, 2000a, 2000b) do. Yet we focus on an outcome of theoretical interest, that is, on a theory-driven narrative as opposed to a history-driven narrative. The resulting model, although still a "just-so" construct, captures a generic dynamic rather than a unique and momentous play in history. It is thus sufficiently abstract to be generalizable and to lead to predictions, some of which may be testable.

2.3.5. Models to Evidence: M → E

As stated, models test, or rather, pretest theories by confirming the logicality of their conjectures. Via modeling, we refine assumptions and establish parametric bounds, through the iterative design process. Formal modeling thus bears the epistemological weight in the process by which we consolidate and extend the body of shared knowledge—by which we expand Popper's "World 3" (e.g., 1972, 1978). This by itself is justification enough to invest some effort in formalizing our arguments. Yet models can generate additional benefits. Specifically, though this is not their primary role, models can also generate useful new ideas for empirical testing.

At this, interpretation, stage, M → E, the initial theory is already refined and likely made more specific. Yet even though more specific, we now attribute to it the power to explain multiple facts that have in common their capacity to meet the revised theory's (i.e., the model's) assumptions. To recap the steps: we refine the original theory, T, as based on some observation. That can hardly be characterized as capturing a regularity, not yet, but it carries such potential if the observations that we use fall into a broader class. A refined theoretical prediction, T′, which is more or less modified from T, has the benefit of ensured *Existence* and *Logicality* under the assumptions that are derived from observation of something in the concrete world (E_j). (We invoke figures 2.2 and 2.3.) Finally, we may venture on to export the logic of T′ to the world of interpretations. That is, only after we model do we generalize explicitly to a larger empirical world (E). As part of such generalization, some form of evidence-based evaluation might be possible, including falsification. Even a successful practical application is a way to obtain empirical confirmation, as long as we have satisfactory criteria to evaluate what constitutes success.[11]

The problem is that both the narrative and the model are sui generis, allowing for no external validity. To the extent to which we pursue a generalizable argument through modeling, external validity is necessary. In the empirical universe, the foundation for critically examining a theory's generalizability is to scrutinize its predictions against as many data points as appropriate and possible. To be able to do that, we want to draw predictions of sufficient generality that more than one observation would meet the constraints on their applicability, creating some possibility of falsification. The fewer the degrees of freedom (i.e., the more assumptions included), the fewer similar observations we would tend to have in each group. In the vein of most similar case comparison, we would need several otherwise similar cases in order to discern the separable effect of the independent variable on which the cases disagree.

When empirically examining the generalizability of a formal argument, we have two potential approaches. If looking to confirm the model's predictions as deterministic, then an appropriate approach would be a small-N research design with very precise case selection criteria. The exact same circumstances in very similar, carefully selected cases should yield the same results. When stated that way, in order to compare a model with evidence, we would need at least several stories where the same structure of social interactions under the same parameters should lead to the same outcome.

Alternatively, we could forego the expectation of determinism and thus loosen our selection criteria. Then we can implement a large-N design and statistical analysis. Whenever case selection is not precise enough to guarantee that the assumptions of the formal model are met exactly, any deterministic aim is lost.

Let us recognize again that empirical testing does not necessarily have to be a part of the T′ → E stage. Generalizing is possible without hypothesis testing. With formal models, our primary output is what Clarke and Primo (2007, 741) delicately label "scientific reasoning" as opposed to scientific research. Rigorous empirical testing is excellent whenever possible, but often direct empirical examination of a model's prediction(s) is simply not feasible. Consider, for example, the Prisoners' Dilemma: the model itself is nicely generalizable, but its findings do not yield themselves to hypothesis testing. The PD's predictions are offset in the empirical universe by the institutional solutions that people implement precisely in order to alleviate the adverse dynamics that the PD predicts.

Some predictions of formal models are inherently nonobservable and nonoperationalizable (e.g., when the predicted behavior is to do nothing). It is important to acknowledge that such nonobservable predictions may emerge, which is normal. On the other hand, the same model may simultaneously generate observable implications, even if only tangential to its main gist, and those can be subjected to empirical testing. When some part of a model's predictions can be tested, that testing validates its underlying theory. Insofar as that theory has at the same time predicted the nonobservables, such nonobservables have now been indirectly confirmed also. We call this process "testing by association." Nonoperationalizable predictions can be viewed as not rejected insofar as their parent theory has not been rejected.

In testing by association, the use of formal modeling—where multiple predictions emerge as logically linked—extends our ability to develop knowledge where direct observation and operationalization are unable to reach. One could argue that formal modeling augments the power of the scientific method to validate the conjectures where otherwise one would have been reduced to speculation. That is to say, modeling can extricate a social scientist from the quandary of not being able to conjecture credibly about nonobservable yet interesting phenomena, like, for instance, not being able to conjecture about party discipline because party discipline is unobservable when it works.

2.4. Single Story, Multiple Narratives, Even More Models

As stated, the formal models developed in this book originate in real-world events as reported in the news. Whereas all stories contain strategic actors and strategic interactions, none is an earthshaking story that has changed the course of world history. We do not treat coverage of, for instance, the assassination of Archduke Franz Ferdinand of Austria in June 1914. Whenever the story is "big," a sole question arises that dwarfs everything else. To continue the example of the Archduke, debate has long centered on the question, what were the origins of World War I?

Most illustrative cases in the social sciences arise from big stories. Throughout human history, people have taken lessons from big stories and the puzzles they pose (cf., e.g., Bates et al. 2000b, esp. 697 on the team's choice of puzzles; Machiavelli 1975 [1532]). The important puzzle and its theoretical resolution in such stories provide stepping-stones to big generalizations.

Big stories, however, do not fully support a diversity of theoretical approaches. They typically direct attention to the "right" important question and so to the "right" model and way of narrating the events. To be sure, we could still ask questions other than the "right" one, but, by virtue of the story's selection, a single set of strategic dynamics would dominate all others in importance, and as a result all other models would appear superfluous.

Where no single interpretation obviously dominates, how many narratives and models can be extracted from a single story? The short answer: many. Based on each story, we proceed along multiple theoretical routes, which means that we end up with multiple renditions (multiple narratives), and potentially even more models, all "summing up" the same piece of reality. Here again lies the epistemological difference between our approach and that of analytic narratives. If, as Elster (2007) argues, models are mechanisms by which events are generated, we present multiple such mechanisms generating each event. Does this mean that the mechanisms thus uncovered are mutually exclusive, and that we need to search for the right one, as the analytic narrative methodology would have it? Not so for us, and this is why we choose to select stories for modeling in an almost arbitrary way. With mundane, everyday events, it is plain to see that there is simply no reason to favor one theoretical waveband to all others beyond the analyst's personal preference. Chapters 4 through 7 illustrate this.

If we have multiple mechanisms, are these multiple aspects of a single underlying mechanism—aspects now visible by projecting an event onto different theoretical planes? Or perhaps what we have in hand is a complex machine—a compound mechanism where simple mechanisms from individual theories are mechanically attached to each other in some engineering sequence in order to produce complex work. With theoretical dimensions expressed as models, sometimes the latter can be the case, and chapter 5 presents this possibility.

Specifically, chapter 5 develops several narratives and models, following queries about different actors in a dilemma. We discover, however, the usefulness of incorporating implications of previous models in the design of subsequent ones, just as they all incorporate the implications of the resolved Tragedy of the Commons. We end up with an assembly line of sorts, where we produce smaller machines that form parts of the large machine and we then put them together.

2.5. Models in the Big Picture

Formal models are a distinct module in social research. They play a fundamental epistemological role in linking theory and observation, enabling a scholar to move between the specific and the general. Because the construction of models involves the process of reconciliation between premises and conjectures, in modeling the pathway between specific and general goes both ways.

We put forward the narrative as a necessary precursor to a model. Narrative is a part of the same epistemological block as the model and occasionally even substitutes for the model. This conception of narrative as an informal variant of a model reflects the predominant practice in political economy, although heretofore it has not been epistemologically articulated.

We argue that in the process of their design, models balance the equation between premises and conjectures. When models are treated as equations, that enables the analyst to revise theory. Indeed, our theories need not be, and are not, carved in stone. Because the model fuses theory with some initial observation and checks theory for logical consistency, our theories undergo improvement through modeling.

Social scientists often treat formal modeling as a methodology alone, yet here we assert models' epistemological place. In setting out the epistemological place of modeling in social research, figure 2.1 traces how models link to evidence, narrative, story, and theory. Figures 2.2 and 2.3 encapsulate how we understand the process by which models test and build theory via the iterative reconciliation of theory and model.

While a model is an epistemological block in social research, how we model is a methodology. Formal modeling is a mathematically rigorous methodology, which permits formal models to fulfill the epistemological role we have elucidated here. The next chapter reviews the characteristics of formal models as a methodology.

THREE

Methodology

Strategy and Practice in Modeling for Substantive Research

The chapter develops four major themes—takeaways, we might say. First, we draw distinctions among types of formal models, such as social choice, cooperative game theory, and noncooperative game theory. Second, we open up and further discuss the toolbox to be used in completing the steps outlined in figures 2.2 and 2.3 in chapter 2. Third, although our intuition in modeling may start from real events, the resulting model is inevitably too abstract to correspond with reality. To align a model with any actual observed context would require a number of additional assumptions unique to that context, and a full match between model and actuality can never be attained. Fourth, whether famous or "workhorse," formal models enable theory development for the purpose of accumulation of knowledge.

3.1. The Primacy of Theory

Whenever a model is designed, it is always based on an underlying theory. The implicit or explicit assumptions made in addressing research questions divulge the adopted theoretical stance, whether the analyst explicitly proclaims it or not. This is highlighted by the fact that different theories inspire different models from the same empirical material. To demonstrate the primacy of theory, subsequent chapters show how the same initial story takes different shapes when different theoretical filters separate out the story's elements. That is, the same story is told with different emphases depending on the theoretical "punchline" to which the narrator subordinates the selection of information items. Extracting multiple narratives, all equally compelling to an unprejudiced reader, from a single account of real-world events means that the starting story itself is relatively unimportant, even mundane, since no historically important lesson is consensually associated with it.

3.1.1. The Distinctiveness of Formal Theory as a Methodology

To show that formal theory is a systematic methodology—even if it can be an inspired creative enterprise at the same time—this book uses mundane stories as the sufficient basis to design models. The path to a model for a scholar begins when she chooses which question will anchor her inquiry and which theoretical elements (i.e., which assumptions for her future model) she will adopt to address the question. In the chapters to come, we make explicit that these assumptions are merely choices. We do so by constructing several alternative paths to multiple, unrelated models on the basis of a single story; and we draw several narratives with different theoretical aims from the same factual material. To craft our narratives, once more, we opt to work with journalistic material that is not strongly biased in favor of a single question or puzzle: we deliberately veer away from "big" stories, and toward everyday ones. Because the stories so selected contain no dominant question, it is relatively easy to see how different people can reach different conclusions from them. To illustrate this step of a *choice* in research design in an intuitive way, we do not commit to any single puzzle but instead show how a single set of events can give rise to a multiplicity of theoretical inquiries.

To emphasize that our story selection is free from preexisting theoretical preferences, we pull news items in media coverage on and around an arbitrary day in the calendar. In addition to following our criteria for unbiased coverage as outlined in chapter 1, we allow for a single theoretical bias in our selection of stories: we require the presence of people and conflict in them, because application of rational choice methodology, which we use, requires individual agency.

Our use of everyday stories serves to highlight that *any* small set of facts where people and conflict are present can inspire a model. The reader might wonder: Why is such a degree of arbitrariness acceptable in choosing stories that inspire the models created in this book? Recall that the discussion of research design in chapter 2 emphasized that the methodology of formal theory is not empirical, unlike in statistics. Whereas formal models use a small set of data as a starting point, they do not rely on any systematic body of evidence. Formal modeling does, however, require an extensive mathematical apparatus.

Developing some background in mathematics would obviously serve one well in both formal and statistical modeling. This is why the same scholars often practice both methodologies. Yet the two methods are tasked with very different duties. The formal method answers a twofold question: Under what conditions does a particular outcome occur? And what can be logically expected to happen under a given set of conditions? Quantitative methods in the social sciences, in turn, address the question: Can a theoretical claim be falsified? This process of hypothesis testing unfolds by means of collecting, analyzing, and making inferences from observational, simulated and experimental data. As we reflect on scientific inquiry, inference in statistical methodology and deductive reasoning in formal modeling might be seen as subordinate when compared to the goal of falsification via hypothesis testing (e.g., Kuhn 1962; Clarke and Primo 2012; Lakatos and Musgrave 1970). Since the reader's willingness to read this book signals her acknowledgment of the usefulness of rigorous logicality in constructing theoretical arguments and delineating their boundaries of applicability, we leave the discussion of empirical method for others to pursue.

3.1.2. The Question as Prism

Because we choose everyday stories, each of our selections has the ingredients for many equally viable formal models. None of them raises just one puzzle or question of predominant importance. Rather, each story can lead to multiple questions. Each question, in turn, serves as a prism, separating what we might imagine as a beam of white light, in the story's elements, and letting some wavelengths through while deflecting others. Exploiting this image further, the narrative—and the model based on it—may be understood as a frequency band, on the frequency of a question.

Thus, we deliberately choose and use stories in order to make a particular methodological point: the feasibility of multiple questions and multiple theoretical directions, as well as the primacy of theory. This contrasts with how a scholar would illustrate her theoretical argument by selecting a story to use as an opening vignette or a piece of supporting evidence in a paper. We also differ from scholars who select cases in a small-N comparative design, who strive for an absence of theoretical bias and work with theory-derived empirical criteria that the cases are intended to satisfy. In choosing everyday stories of people and conflict, we create a level playing field for different theoretical angles, which lead to multiple narratives, multiple questions, and multiple models (cf. Page 2018). In contrast with the analytic narratives approach (Bates et al. 1998, 2000a, 2000b), we demonstrate the plausibility of multiple theoretical perspectives, crystallized in models, all inspired from a single real-world story. Even before the model is designed, the story gives rise to theoretical arguments encapsulated in narratives, which some might also call colloquially stories, as in "What story are you telling?" This colloquial reference recognizes narration as a form of presenting a theoretical argument.

Recall that figure 2.1 draws attention to the two stages in the research process where formal modeling connects with evidence. We are now discussing the first such connection, reaching from a single observation to revised theory, T'. This connection proceeds through constructing a formal model. Upon completing the model, we would be able to build the second connection: there might be testable propositions to subject to systematic empirical evaluation. A story, understood as a minimally processed set of facts, serves as an isolated datum that provides a starting point for modeling a theoretical argument.

3.2. The Technology of Modeling: Types of Formal Models

All formal models discussed and designed in this book are *economic* models, meaning that they are all founded on the broad and general premises of microeconomic theory. These are, first, methodological individualism and, second, rationality. Methodological individualism is the premise that outcomes are brought about by choices made by *individuals*. Thus, to explain outcomes, we first need to explain or understand individual behavior. The second, individual rationality, means that actors (individuals, conforming to the first) have goals based on their ability to formulate preferences over alternatives, and act purposefully, that is, in such ways as to attain their goals. Individual rationality stipulates, more specifically, that actors have complete and *transitive* preference orderings over the possible outcomes (i.e., over the feasible alternatives),[1] and possibly well-defined utility functions over the space of the outcomes (Myerson 2013). In principle, since these microfoundations are no more than a special set of assumptions, one could replace those assumptions with some alternative and build up alternative mathematical structures, which would also technically be formal models. We do not explore those directions here since we are not aware of formal work done in political science on premises that are alternative to microeconomics.

What types of economic models are there? And is a scholar required to choose a particular type for a given project? Addressing the second question briefly, the answer is no: a scholar can implement whichever method she prefers, although she may find that she attains less purchase with some methods as opposed to others in a given set of circumstances. Now taking up types of models, we briefly discuss some of the chief possibilities: the utility maximization approach, decision theory, social choice theory, cooperative game theory, and noncooperative game theory. The order in which we have just listed these methods does not indicate their relative popularity, legitimacy, or frequency of use. Furthermore, some models can be reasonably filed under more than one such methodology. Utility maximization, for example, is present in all economic models of behavior—cooperative game theory, decision-theoretic, and noncooperative theoretic—and arguably is implied in social choice theory as well. The reverse is not true, however, and a utility maximization model does not need to be any of those others listed; setting up the utility functions of the players can be a step in research design even without solving a decision-theoretic or a game-theoretic problem.

Occasionally, when game-theoretic models are discussed, we hear a question: Should I go for a game-theoretic or an economic model? This question, well intentioned though it may be, is based on a misunderstanding. What the modeler probably has in mind is utility analysis without an additional layer of strategic analysis. Another type of economic model without game theory is decision theory, a very close relative of game theory. Both decision theory and utility analysis can parameterize an actor's own moves as determinants of that actor's utility. Decision theory is distinct from utility analysis in that it takes into consideration the dimension of time, dealing with the sequences of moves of a player or players. We now look at each type of model in greater detail.

3.2.1. Expected Utility Maximization

All rational choice models make the fundamental assumptions of methodological individualism and individual rationality, as defined above. While there exist rational choice alternatives to the premise that to pursue individual goals means to maximize one's expected utilities (e.g., the maximin solution concept), we do not address such variants here, and refer interested readers instead to, for example, Ferejohn and Fiorina 1974. What we do address, though, is an extremely large kingdom of (micro)economic models, all of which use utility maximization.

Models of *utility maximization* do specifically assume that we have some knowledge of the utility function of the actor. Whereas *cardinal* utility refers to satisfaction that can be expressed numerically, *ordinal* utility refers to rankings of satisfaction that, though ordered, has no numerical expression. *Intensity* of preferences denotes the willingness to make trade-offs between alternatives. Be they cardinal or ordinal, utilites can be affected by the actor's own behavior as well as outside influences, and thus we have an idea of what benefits or hurts the individual. Consistent with the rationality assumption, we further assume that whenever an individual would make choices, she would do so in ways that affect her utility positively.

All of the methods that we survey are in fact variants of utility maximization. A consumer who works with a fixed budget, compares the prices of two goods, and decides how much to buy of each good on the market is engaging in utility maximization, but is making a specific decision out of a fixed set of available choices. We thus would label that model as decision-theoretic. A voter in a democratic election compares the costs of voting (e.g., spending time in gathering information) against its benefits (casting the decisive vote), and decides not to vote. The paradox of voting (e.g., Downs 1957; Fedderson 2004; Ferejohn and Fiorina 1974; for a recent discussion, see Medina 2013) is that turnout rates are always much higher than zero! Again, her utility is maximized. Yet the corresponding model is either game-theoretic or decision-theoretic, depending on whether or not others' behavior is explicitly taken into account in the model as a reason why her vote will not affect the outcome of the election.

What we label here as basic utility maximization models, then, is a class within the larger kingdom of utility maximization models. The class we examine does not require those additional assumptions that define other classes within the same kingdom. In other words, we categorize utility maximization models as a sort of residual category—when the model in question cannot be viewed as, say, utility maximization *and* decision theory, or utility maximization *and* social choice, or utility maximization *and* game theory. (For research using the utility maximization approach thus defined, see, e.g., Adams, Merrill, and Grofman 2005; Akcinaroglu and Radziszewski 2013; Heller and Shvetsova 2016; Mershon and Shvetsova 2011, 2013a, 2013b). As we proceed in our survey of approaches, moving from utility maximization to other methods, we add premises and thus restrict models in more classes.

3.2.2. Decision Theory

Decision theory models how agents engage in expected utility maximization, taking into account uncertainty, and it does not necessarily require the analytical justification of beliefs arising in connection with other agents' expected strategies.[2] While both social choice and game theory can be considered subsets of decision theory (insofar as the latter follows the rationality criterion), we place them in separate classes of models, because they require further, additional assumptions. What we refer to here as decision-theoretic models are a class of models that cannot be attributed to either social choice or game theory, yet generate a deterministic prediction of an actor's choice. In decision theory, utility maximization leads to making a specific choice out of a well-defined set of choices by a single actor.

The foundational *The Calculus of Consent* (Buchanan and Tullock 1962) uses decision-theoretic models. Other scholarship using decision theory includes such broadly influential works as Alt and Woolley 1982, Axelrod 2015, and Gartner and Regan 1996. Just because decision theory is identified here as "not social choice or game theory" does not mean these works are any less technically demanding; see, for example, Kollman, Miller, and Page 1992 and Page 1999.

3.2.3. Social Choice Theory

Social choice theory historically and methodologically precedes game theory: it is more "basic" and far older than both cooperative and noncooperative games. Social choice does not go as far as to presume to know the exact utility functions of the players. Instead, it limits itself to assuming basic individual rationality—the players' ability to compare specific pairs of feasible alternatives with some assumed level of consistency. It then aggregates individual preferences into social choices via a range of aggregation rules, discovering most importantly that social choices based on well-behaving individual preferences do not inherit, from the individual level, the same well-behaving properties. Crucially, social choices have the propensity to violate the property of transitivity (Arrow 1951). Condorcet (1788) and Arrow (1951) figure among the masters who have demonstrated nontransitivity at the aggregate, social level despite individual-level transitivity (cf. McLean and Urken 1995). The social choice approach permits analysis of the characteristics of preference-aggregation procedures that may arise when those procedures are applied to groups or societies with preference structures characterized by conflict.

In constructing a social choice model, we need to specify a finite set of feasible alternatives, and for each player specify her preference orderings (i.e., how she "ranks" the alternatives in her preferences—her ordinal utility function). In place of a solution concept, social choice models use prespecified sets of axioms, requiring that those be simultaneously satisfied. In more specifically aimed models, we make fixed assumptions about the decision rule and the ability of the players to anticipate the consequences of their present choices for the future, when subsequent choices would arise. Social choice models may also let players evaluate alternatives via a numerical ranking, with some counting rule then applied to determine the winner. A Borda count is one example. No matter the details of the social choice model, our question is always about the properties of social choices—given what we know about the preferences of the rational individuals who participate in them. In political science, for example, highly influential accounts of the formation of political parties use social choice theory (Aldrich 2011 [1995]). The definitive texts in social choice theory include Arrow 1951, Fishburn 2015, Plott 1976, Saari 2012, and Sen 1986. For applied social choice research, see, for example, Chandra 2001; Riker 1982; Saari and Sieberg 2004; Shepsle and Weingast 1982, among many others.

3.2.4. Cooperative Game Theory

Cooperative game theory defines its main solution concept as a set of outcomes, and tests the stability of a choice of outcome by the criterion of the unavailability of decisive coalition-supported alternatives. This starkly contrasts with the criterion of the irrationality of individual defections (as in noncooperative game theory). The question of cooperative game theory is of a material type: What will the outcome be? The choices before players in cooperative game theory, then, are not really strategies: rather, players choose which coalition—which group advocating which outcome—to join when confronted with a given status quo. In most traditional cooperative models, players do not really make any moves. Instead, a scholar sifts through the players' *hypothetical* moves across various coalitions as she searches for policy locations that would be immune to any such moves. Then, as her solution concept indicates, she assumes that the outcomes (thus, the implied actions) would land somewhere in these locations.

Hence cooperative games require that we specify players' policy ideals (their most-preferred policies), their utility functions (or the shapes of their indifference curves in the policy space), players' voting weights, the status quo or reversion policy location, and the decision rule, which can be majority rule or some qualified majority (i.e., some supermajority) and which defines the minimal required size of a decisive coalition. Furthermore, the scholar usually commits to a specific variation on the main solution concept. There are several such variations that cooperative game theorists use, but the fundamental original solution concept is the *core*, defined as the set of outcomes that cannot be defeated by any decisive coalition.[3] Researchers using cooperative game theory models have obtained fundamental results on the instability of decision making in multidimensional policy space (McKelvey 1976, 1979; McKelvey and Schofield 1986; 1987; Plott 1967; Schofield 1983, 1986; Schofield, Grofman, and Feld 1988). Applied cooperative game theoretic work encompasses, for example, Epstein and Mershon 1996; Heller 2001b; Laver and Shepsle 1990; Laver and Schofield 1990; Mershon 2002; Mershon and Shvetsova 2014; Miller and Schofield 2003; Monroe and Robinson 2008; Schofield 1993b, 2003, 2004; Schofield and Laver 1985.

Scholars have developed a number of mathematical results that establish some equivalences and correspondences between solution concepts in social choice, cooperative, and noncooperative game theory. In practice, though, applied work rarely invokes such equivalences, and a substantive model falls into one of these categories, depending on the analyst's choice of premises.

3.2.5. Noncooperative Game Theory

The underlying premises of noncooperative game theory are individual rationality and common knowledge. The assumption of individual rationality again indicates that these models are a subclass in the class of economic models. The assumption of common knowledge, however, marks a new addition to the standard economic setup and is the distinguishing premise of noncooperative game theory. Common knowledge means that actors are aware of each other's goals and of each other's purposeful actions to achieve those goals, which means that they have beliefs about the behavior of other actors in the game and they take that into consideration when determining their own behavior.

The question that a noncooperative game answers is behavioral: What will actors do? The solution concept is the Nash equilibrium, which is defined as a combination—a profile—of players' strategies. Specifically, these strategies are such that no individuals can improve their utility by unilaterally altering their own play while the rest continue as they intended.

So defined, the Nash equilibrium maximizes a player's utility, conditionally on the fixed behaviors of others.

A reader might encounter a noncooperative game that seems to be an application of game theory yet not so much of economics. This perception might emerge because noncooperative models can sometimes be so spare that they can be analyzed in straightforward ways without resorting to an explicit calculus exercise in constrained maximization. Such simplicity in our view is a desirable quality, but solutions to this game still amount to utility maximization under the constraint of an additional assumption (common knowledge) and thus remain in the realm of microeconomics. Nash equilibria equivalence with utility maximization is a fundamental result contained in the expected utility maximization theorem (in continuous strategy spaces with some general restrictions on the actors' utility functions; see Myerson 2013).

While the Nash equilibrium corresponds to the conditional maxima of *individual* utilities, it only occasionally coincides with an overall efficient outcome or with maximal total welfare. This is because strategic players do not pursue either of those latter goals, as per methodological individualism. Instead, they pursue their own individual welfare, and through their behavior they are capable of undermining the collective good. Yet once again for a strategic individual who confronts other strategic individuals, the choice of a best strategy constitutes utility maximization. We emphasize that noncooperative games look for solutions in terms of players' behavior, and view the choices before actors as choices of strategy, not choices of outcomes.

As one instance of research employing noncooperative game theory, consider the model of Alt, Calvert, and Humes (1988), which incorporates incomplete information, repeated play, and reputation in order to shed light on Saudi Arabia's strategy in international oil markets. Other scholarship that uses noncooperative games includes Akcinaroglu and Radziszewski 2005; Ansolabehere and Snyder 2000; Bednar and Page 2007; Bianco and Bates 1990; Filippov, Ordeshook, and Shvetsova 2004; Gerber 1996; Groseclose and McCarty 2001; Heller 1997, 2001b; Heller and Sieberg 2008, 2010; Huber and Lupia 2001; Huber and McCarty 2004; Kinsey and Shvetsova 2008; Lupia and Strøm 1995; Palfrey and Rosenthal 1985; and Snyder and Ting 2002.

3.3. Knowing the Steps and Stepping in Sequence

The methods just surveyed differ in their assumptions, solution concepts, and other components. We now build on this inventory to make an explicit comparison of their fundamentals. In building a formal model, whenever in doubt we can always fall back on identifying the requisite items of which our model *must* consist in accordance with the formal rules. While different types of models have somewhat different requisite components, these lists of items are finite and intuitive. Thinking about the components that need to go into various types of formal models is thus akin to knowing the steps in the modeling enterprise. Fitting these components together, in alignment with story, narrative, and theory, with any needed adjustments (recalling figures 2.2 and 2.3 again), is akin to stepping in sequence.

Chapters 4 through 7 use basic utility maximization, decision-theoretic, social choice, cooperative, and noncooperative game-theoretic approaches to modeling. All five types are also found in the political and social science literature, and each type is responsible for fundamental theoretical results on which all students of human behavior rely in one way or another.

Table 3.1 arrays all five types of formal models and lists their components. The leftmost type in table 3.1, basic utility maximization, is the simplest. Appearing on the far right, the noncooperative game-theoretic model has components that are a superset of those of the decision-theoretic, social choice, and utility-maximization models; this superset is nicely contrasted with the components of a cooperative model.

Below we outline in some detail the main elements of a noncooperative game, even though the previous section already gave a minimal overview. We do so because in noncooperative games we find most of the elements that appear elsewhere in Table 3.1—in other types of models. Noncooperative games thus contain, in addition to their strategy-related elements, the elements that constitute the models of other surveyed types. Readers will easily draw comparisons with other modeling approaches in table 3.1.

The first item in the list of components is *actors*. Not everyone mentioned in a story must become an actor in a model. Even at the stage of the narrative, some individuals and some actions may be omitted. Others may remain, but merely as contributors to the final gains or losses, as they affect those, but in fully predictable fashion. That only some individuals mentioned become actors is good, because analytical complexity jumps with every addition to this set. Actors are those "individuals" who, in noncooperative games, will behave strategically, that is, in anticipation of others' behaviors. For the practical purpose of solving a game, two actors are enough, three are many, and four are almost too many: four may introduce too much analytic complexity. To appreciate the potential for complexity, bear in mind that each actor may have many contingencies in which she gets to move.

This brings us to the next component: *actions* that are available to each actor. Actions are specific to an individual actor and to the contingency (in the noncooperative game—specific to an information set) in which that actor gets to act (move).

Information sets are contingencies in the game for the purpose of an actor's moves (actions). Moreover, information sets are the contingencies for action as the actor who is about to move perceives them. This component accounts for the possibility that an actor may not quite know where exactly in the game she is when it is her turn to move—what type of an opponent she is facing, which exact rules apply, whether certain events already took place. Information sets also capture the circumstance where she has to take action before the actions of others are revealed. In all these

TABLE 3.1. Comparison of components across types of formal models

Model Components	Types of Formal Models				
	Utility maximization	Decision Theory	Social choice theory	Cooperative game theory	Non-cooperative game theory
Actors	One	One	Multiple	Multiple	Multiple
Actions	No	Multiple in each information set	Only in models of voting	No	Multiple in each information set
Information	Uncertainty	Incomplete and imperfect (uncertainty)	Complete	Complete	Incomplete and imperfect (uncertainty)
Outcomes	Not necessarily	Yes	Yes	Yes	Yes
Preferences over outcomes	Cardinal utility functions	Cardinal or ordinal utility functions	Preference orderings (ordinal utility)	Cardinal utility functions	Cardinal or ordinal utility functions
Beliefs	Probability weights on states of world	Probability weights on states of world	No	No	Probability weights on states of world + on actions of others
Strategies	Constrained maximization	Choice of action	Choice of an outcome	No	Choice of plans of actions
Solution concept	Constrained maximization	Utility maximization	Axiomatically defined for each model	Outcomes are stable (core: unbeatable)	Nash equilibrium

circumstances, the actor must act while not knowing the actual state of the world (i.e., the actual environment facing her). Then again, there are information sets that depict perfect clarity, contingencies when none of the above applies and the player knows exactly what is going on and where she is in the game (these are singleton information sets).

Where information is imperfect (that is, information sets contain multiple states of the world), choosing action requires holding *beliefs* about the relative likelihood of possible states of the world. Beliefs are probability weights on alternative states of the world within an information set. They are also a component of the game.

Whichever way the actors elect to move in a game, the end result is some *outcome*. An outcome is the state of the world after the game is over, and therefore in itself is the same for all players; it is a good of sorts that the game has produced. Different actions would lead to outcomes that may be different or similar. Different combinations of individual moves may produce equivalent outcomes, but each produces only one outcome. In all, there are usually multiple outcomes that can feasibly be reached in a game. Even though we may be primarily interested in one or a few outcomes in the game, we need to be careful to account for all feasible outcomes—for every state of the world that might result from the specified players' feasible actions. Outcomes are yet another component of the game.

The outcome is where the game ends. It is also the reason the game is played. The conflict of preferences over the outcomes provides the source of conflict in the game. Though each actor contributes to the result individually by taking actions when it is her turn, the path to the outcome is laid by the actions of all actors. This does not mean, however, that identical outcomes cannot result from different combinations of individual moves. Depending on a game, this can happen.

The *preferences* that individual actors hold over the feasible outcomes of the game are a distinctive component of the game as well, and are what motivate actors' choices. It is up to the modeler to decide whether she needs to assess the intensity of actors' preferences or it would suffice to merely specify the order in which outcomes are valued by them. Depending on that decision, the analyst will make assumptions about either the cardinal or the ordinal utility functions of the actors over the outcomes in a game. Actors' *utility over feasible outcomes* is another component of a noncooperative game that must be specified.

Everything in game theory, and also everything that we call policies, in essence amounts to the outcomes shared by all those involved. Again, an outcome occurs one for all. The same president is elected into office for every voter no matter which way she voted, the same healthcare reform is enacted for every citizen, the same tax code is imposed on everyone, and so forth. What differs is how actors relate to those outcomes—how much they like or dislike them. That is the domain of preferences, or utilities, and it is vital: preferences and utilities are essential for solving the game. Making assumptions about the preferences of actors over outcomes is where the scholar exercises the greatest degree of judgment and applies her theoretical bias.

Accounting for all feasible outcomes and making reasonable assumptions about the actors' utilities from those is, perhaps, the most difficult step in drafting a model. It requires patience in going over all of the feasible end results of the game, no matter how unlikely the scholar believes the game to end there. This is so because strategic behavior is driven by an awareness of the paths that the actors wish to avoid as much as of those that the actors prefer to travel. This work also requires making decisions about what relative values to assign to an actor in the model for outcomes that may not be easily compared. While sometimes the preference is "obvious," and the social scientist can turn to conventional wisdom, scholarly and otherwise (e.g., less time in jail is better than more time in jail), at other times the analyst faces a tough judgment call. When that happens, she may argue that a player prefers one of the outcomes by relying on her prior knowledge; the cost of such an approach is opening herself to possible criticism for her choice of assumptions. Alternatively, she may leave the matter unresolved, allow for the possibilities of either of the two outcomes being preferred by a player, and analyze the game separately for each of those assumptions.

All the components specified so far will factor into players' choices of *strategies*. The set of strategies available to each of the players is the next component. Strategies are most commonly defined as plans of action for all contingencies. That means that each of the strategies includes provisions for how to act in each of the information sets of a player. Again, it is essential that all such feasible plans are taken into account, regardless of how clever or foolish they may seem: any strategy is clever or foolish only in the context of the strategies of one's opponent, and in solving a game, we pit strategies against strategies.

The final component of the game is the *solution concept*. The solution concept is what we *assume* will constitute the prediction in a game. A solution concept describes the necessary and sufficient characteristics of the thing that we are willing to embrace as an answer. In that, it is a set of assumptions, sometimes labeled as assumed conditions, that need to be satisfied. The fundamental solution concept of noncooperative game theory is Nash equilibrium. The conditions existing in this solution are such that each actor plays the strategy that gives her a maximal utility for the combination of strategies that she believes the others are playing.

There are two logically possible ways to portray this connection between what actors do and how the game ends. One is to take an outcome and list all possible combinations of moves that result in it. The other is to look at each combination of moves and, to each combination, attach the outcome that it brings about. Since in noncooperative game theory we are ultimately interested in behavior—in what actors choose to *do*, what strategies they will choose to play—it will help us to follow the second option. For this reason, in noncooperative game theory we (1) take each combination of moves and (2) evaluate what the outcome would be if these moves were to take place in a game.

As indicated, the components of noncooperative game theory just reviewed are found across other types of models, although not all of them are present at the same time. For instance, a single actor figures in utility maximization and decision theory, as shown on the top row of table 3.1, whereas multiple actors feature in social choice theory, just as they do in both cooperative game theory and noncooperative game theory. This contrast alone highlights that choice of method is consequential for a model's predictions, just as behavior of an isolated human being would differ from human behavior in a social interaction.

As an example of difference in method leading to difference in predictions, consider Boix (1999) and Kinsey and Shvetsova (2008), who model the same social phenomenon—franchise expansion in European democracies at the turn of the twentieth century—but do so in different ways. In particular, Boix (1999) uses a decision-theoretic model, whereas Kinsey and Shvetsova (2008) design a noncooperative game. Given that distinction, the two works reach different conclusions.[4]

3.4. The Generality of a Model

A widely heard saying has it that "a model is only as good as its assumptions." In what sense might this be true, and can a counterclaim be made? Assumptions provide the starting point of a theory, from which everything else flows, all the way to conjectures and predictions. In a formal model, the coupling between this first link, assumptions, and each subsequent link in the chain of reasoning is so tight and firm that weak assumptions weaken the entire chain. A model cannot escape its weak foundations.

But what makes assumptions weak? A mere lack of realism in assumptions does not necessarily undermine a model, as the Prisoner's Dilemma illustrates so well. On the other hand, a model would suffer fatally from inconsistency in its assumptions. Overall, the modeler has extraordinary freedom in choosing the assumptions for her model; this is the only juncture at which she chooses, for the rest falls out in a logical—deterministic—way. What should the modeler consider as she weighs her options? Assumptions, clearly, demand a closer look.

3.4.1. Assumptions: The Fewer the Better

Should one make many assumptions or few? When the question is posed in this way, it seems rhetorical: of course, the fewer assumptions the better, for the model will be then more general. Would anyone argue against that? Indeed, she might. Consider a statistical model with multiple controls, introduced alongside the independent variables tapping hypothesized influences. The object is to reduce unexplained variation; if there might be influences out there in the world that, beyond the posited ones, might be able to account for variations in the dependent variable, most researchers would be inclined to account for the possibilities with controls in their estimations. With each such addition, they de facto make at least one extra assumption.

The difference between formal and statistical models is that in a formal model we do not want to explain most of the variance. Instead, we want to maximize the model's generality—its applicability to as diverse a set of cases as possible. Elsewhere, we address this difference between formal models and statistical models. Generality is valuable in a formal model, as it is in a theory. The problem is that researchers are often forced to refine a theoretical argument by incorporating a range of assumptions into their models, for example, when that is necessary to make their models work, as in the theory-model reconciliation algorithm depicted in figures 2.2 and 2.3.

Thus, we advocate making as few assumptions as possible and streamlining models as much as possible. We often see a model presented in as simple a form as possible, with the analyst then taking subsequent steps, varying the simplest, "baseline" model by sequentially altering or adding assumptions. Indeed, the well-known Occam's razor (after William of Ockham, of the medieval era) holds that the best explanation, out of several possible ones, is that with the fewest assumptions. By this principle, a theoretical model should be "shaved down" to its essential assumptions, and nothing more.

As we work from a theory-motivated narrative, we start with the simplest model possible and proceed to more elaborate versions only if necessary. We are going to regard the first draft of a model as but a study for our future masterpiece, most likely to be redrawn or even discarded. Figures 2.2 and 2.3 in chapter 2, with their iterated modifications of assumptions in the modeling process, illustrate the benefits of using the sparest model as an initial baseline. Whether or not complicating the model is warranted will become evident in time. Through the process of modeling, we will discover whether or not our questions are answered by a relatively simple model or require more elaborate versions before we can establish the *Existence* and the *Logicality* of the argument. We have guidance on the specifics of this process in figures 2.2 and 2.3, which outline the steps to follow in case the initial model fails to generate the equivalent of the expected theoretical prediction.

Hence epistemology dictates that we should start as simply as possible. If we lose something interesting due to ruthless simplification, we will surely notice, since the model will not generate the dynamics that we were hoping to capture. If we perceive such a loss, we will add what is necessary: we should plan to address that possibility later, and not worry at the start. We must start somewhere, and it is best to begin with a spare model.

We both expect and exploit the process of discovery. When working on an evolving model, the model's moving parts come into focus. Once the basic model is set up, we will see better what (added) assumptions are needed to induce changes in actors' behavior. In mathematical terms, we can perform the comparative statics on each of the parameters of the model. We will also see how adding (or removing) some of the assumptions, or even including additional players, can turn around the model's predictions.

3.4.2. The Prisoners' Dilemma: Illustrious and Illustrative

Every scholar wants to come up with a model as succinct and powerful as, for example, the Prisoners' Dilemma. She wants to design a useful model, a model that will be cited in the literature by its proper name, like MAD (Mutual Assured Destruction), the Centipede Game, and so forth. Or her ideal model may acquire a proper name that incorporates the name of its authors, such as the "Shepsle-Weingast structure-induced equilibrium" (1981) or the "Baron-Ferejohn bargaining model" (1989). Why do some models and not others become famous in this way? What is it exactly that the scholarly market rewards via the operation of its Invisible Hand? (Aha! We bump up against yet another proper-name model!)

Models win fame when they summarize, convey, and simplify—maximally—a fairly complex logical argument. A model gains renown as more and more researchers view its argument as broadly relevant—that is, as scholars more and more often need to invoke the logic and take a short-cut in doing so. In this way, we can see a formal model as a programming macro: when the macro is frequently needed and used, its name becomes memorable and associated directly with the logical operation(s) it contains. As analysts repeatedly find it useful to deploy such a model in a variety of contexts, its power and range become manifest. The interconnectedness across scholarly works via the shared use of such a model further contributes to the accumulation of theoretical knowledge.

The Prisoners' Dilemma is likely the formal model with the greatest fame. Despite the PD's fame, every article and book that uses it explains what the game looks like, as indeed we ourselves do in chapter 5. It is always a pleasure to read—and write—about the setup and the prediction of the Prisoners' Dilemma and about the simple way in which the result unfolds with inescapable logic. The PD blends the political and the economic. The game is about politics because all of its players share not only the outcome but also their disappointment in the outcome—the loss in utility from what could have been a Pareto-efficient turn of events, which cannot materialize, however, unless and until something about the game changes. The game is economic in that its players are strict utility maximizers: each one of them does what is efficient for each as an individual. It has been said that the PD captures the fundamental logic of the causes and perils of mutual exploitation, and that it conveys the mutual fears of such exploitation in a society or any group of people in society. It has also been said that the PD explains the rationale for empowering the state with its coercive powers. The PD is one of the most generic ways of looking at a society's problems.

The example of the PD propels the question again to the forefront: What is it that makes for a model's success? As the PD suggests, it is not a model's ability to photograph reality that—always or, possibly, ever—makes it popular. A model acquires renown to the degree that its simplicity extends adaptation.

And yet, precisely because it is so generic, anyone would be hard pressed to find an empirical case of what *is*, unquestionably, the Prisoners' Dilemma in real life. How closely does the behavior of people in real-life situations actually resemble that of the prisoners in this game? Critics of the PD would counter that real people do not act like the prisoners caught in the model's dilemma. The critics would emphasize, using the particulars of each specific case, that the PD's setup overlooks important traits not only of individuals but also of institutions. The consequence of such omissions is profound: taking into account real-world characteristics of individuals and institutions affects what players get out of their interactions.

Start with the fact that, in the real world, players are interconnected. If, say, they need to build a community center, some citizens might belong to the local neighborhood association. As they participate in the association, citizens talk about the project as a boon to the community. They also discuss how good it is that everyone in the community will benefit if each chips in his or her bit to finance the construction. Such a conversation might turn a public goods provision problem into a charitable fund-raiser, where donors expect no or few tangible benefits to accrue to them personally. A few enterprising friends in the association decide to organize block parties where neighbors can contribute, as they enjoy the festivities, to the financing of the project. Naturally, the organizers of the block parties ensure that each party hosts speakers who spotlight advantages of the community center; speaking at several parties, for example, are parents whose children will participate in activities there without having to travel outside the community. These forms of communication and persuasion have no place in the PD as an abstract model. In fact, if made a part of the formal structure, these would modify the game enough to have the "dilemma" resolved, by including positive externalities into individuals' utility functions.

Detailed examples of this sort can again and again refute the argument that defection is the individually optimal behavior. Even societies on a larger scale defy PD-driven expectations and manage to beat its inefficient prediction (see Ostrom 1990). And yet we continue to rely on the model—the logic—of the Prisoners' Dilemma with a fervor bordering on the religious; tellingly, the relevant Google Scholar search returned 243,000 results as of late September 2018.

Why? The role of an abstract logical model in understanding a concrete situation, or, even more so, of many concrete situations, is to parse out what is essential. The abstract model generalizes by means of simplifying the complexity. And the more successfully it simplifies, the more powerfully it generalizes. As it does so, the more likely it is to acquire fame. In the way that the PD is illustrious, it is also illustrative, then. This famous model bears witness to the fact that realism in assumptions has precious little to do with a model's elegance, power, renown, and reuse. What counts is not the model's realism but rather its capacity to expose the essentials of a complex situation.

3.5. Conclusion

As we stress at the outset, the main question for us is the epistemologi-cal place that modeling has in the path to knowledge. The epistemologi-cal role of modeling and the many benefits that models yield to research are all, however, premised on the rigor and mathematical clarity of the method. The nature of the method, its logicality and replicability, are what make modeling a distinct contributor to theory development. The other aspect of the method—its specificity, in the sense that we have to be fully transparent about the assumptions of a model—allows modeling to be a venue for testing theories. The demand for specificity of assumptions forces a model to draw from both sides in the epistemological chain, fusing theory with observation. Through that fusion, theoretical conjectures are clarified and tested.

Thus, it is the method that enables the epistemological function. Clar-ity and consistency of method *are* the necessary condition for the fulfill-ment of the epistemological purpose of this research module. This is why we in this chapter have devoted about as much effort to separating the distinct approaches in formal theory, stressing differences among them, as we have given to their descriptions. It is less important which method is applied than that the method chosen is cleanly (clearly and consistently) executed. Our inventory of the different types of formal models and their constituent elements shows that these are but resources to be used to build, develop, and validate theory.[5]

The renowned philosopher of science Nancy Cartwright (1999) argues that conceiving of modeling as a "vending machine" is inadequate as a depiction of how scientists (in particular, physicists and economists) actu-ally use models. The stance we take here both concurs and disagrees with Cartwright's (1999) position. We insist that the epistemological function of modeling requires the enforcement of clarity within modeling's meth-odological toolbox. We claim that order and clarity in the toolbox are nec-essary precursors to models' epistemological function. In other words, a fusion of potato chips with a candy bar just would not serve the purpose. In this way, the vending machine analogy is actually appropriate, in that a clear choice must be made in terms of the method for the model, which is the same as the clear, consistent choice of the model's basic assumptions.

Yet modeling starts, not ends, with the choice of a model's type—with the choice of its basic assumptions. It remains just a type, not a particular model, until we add those further assumptions that generate the specificity just mentioned and that come from both the initial theory and the observation. This is where the "vending machine" analogy fails: the chosen tools cannot be used automatically or passively. Completed models do not come prepackaged and ready for immediate consumption in ways akin to putting in money, pressing a button, and waiting for the vending machine to eject the preferred result. Instead, a model must be designed to a purpose. And that purpose is to refine a specific theory in view of observations that seem like a good fit. The process of design is complex; it goes through the analyst's decision making about the initial setup as well as subsequent steps in reconciliation of theory and model. Illustrations of such decision making by the analyst in chapters 4 through 7 show that modeling does constitute a creative enterprise. Indeed, creative steps are required within a systematically structured process.

The chapters that follow illustrate how to design formal models afresh. If some model has already been created, then whatever theoretical argument it addresses has gone through the process of testing and amending. Unless the purpose were to refute the result of the prior off-the-shelf model (in which case the analyst would argue that it was inappropriate to the task and needed to be redesigned), an off-the-shelf model is already solved. For a new theory, a new model must be made. If that novel model ends up looking similar to the first, that similarity needs to emerge independently, through the process of modeling. The epistemological sequence $(T \rightarrow M \rightarrow T')$ cannot be fulfilled without the design process.

FOUR

Community Effort

Rewarding or Requiring a Reward?

We now proceed with our first application of the algorithm for narratives and models. This chapter uses one type of model, the utility maximization approach. The chapter opens by relaying in full a single news item, and then draws out multiple narratives based on alternative readings of the story. Each narrative gives rise to separate theoretical arguments. We underscore that in our progression to model design in this chapter, as in the chapters to come, we implement the schema in figure 2.1 by showing how to start with a story, fashion a narrative, and build a model.[1]

This chapter's news story originates in the South African town of East London, on the coast of the Indian Ocean, which serves as the seat of the large Buffalo City metropolitan municipality. In 2014, East London's daily newspaper covered at length a local Xhosa chief's initiative to reward community-oriented efforts made by citizens under his jurisdiction. The *Daily Dispatch* entitled its story, "Mbhashe Chief to Reward Trailblazers." This item is close in time to the date for stories in the other chapters, and conforms fully to the criteria for an unbiased news story laid out in chapter 1. Note that it speaks in multiple voices, is peppered with fact after fact, and is not in any obvious way tinged with the journalist's interpretations of opinions and events. This story so nearly embodies our ideal in this regard that we use it, despite its publication some months earlier than our chosen date.

We invite readers, as they read the full news story, to begin to apply the algorithm for narratives and models by imagining a few of the multiple narratives that might be extracted here. Going further, the next step will be to identify multiple theoretical arguments, each based on a separate narrative, and therefore obtain elements for multiple formal models. This last step is the most challenging, yet the earlier step of drawing out narratives is equally crucial.

Story:

Mbhashe Chief to Reward Trailblazers

Gugu Phandle

DispatchLIVE | 2014-06-06[2]

IMPRESSED by his hard-working nation, that subscribe to the notion of *Vukuzenzele* (wake up and do it for yourself), senior traditional leader and head of Mbhashe Traditional Council, *Nkosi* [chief] Xhanti Sigcawu, will reward those individuals who go the extra mile.

Thirty-six individuals, who excel in anything from farming to healthcare, will be rewarded with a trophy and have their names inscribed on an honours board for posterity at a special ceremony today. This gesture signals the beginning of an annual awards ceremony—the inaugural Mbhashe Traditional Council Awards—at the Mbhashe Royal Village Lodge on the banks of the Mbhashe River, with Sigcawu saying there could eventually be some additional benefits attached to these awards. "We have decided to establish these accolades to honour and celebrate talent and dedication within our locality to motivate our people, so they can see the importance of commitment to development and excellence," said Sigcawu, who is the uncle of the AmaXhosa King, Mpendulo Sigcawu.

The awards ceremony is organised in conjunction with the Mbhashe Development Authority Trust and the Culture, Arts, Tourism, Hospitality and Sports Sector Training Authorities.

Sigcawu, whose area of jurisdiction spreads west of the Mbhashe River to include the administration areas of Candu, Nywarha, Bholotwa, Lotha, Nqabarha and Ludondolo, said his people's commitment, dedication and willingness to help others made him a proud leader.

"In most villages in other areas of the Eastern Cape land lies idle and the quantity of livestock has decreased. Even the livestock that is kept, is often of poor quality. However, my people are working the land, feeding themselves and others. The quality of livestock has improved tremendously over the years. These days Nguni cattle, Dohne sheep and other quality mutton stock are bred," said Sigcawu.

Those rewarded today include farmers (grain farming, livestock keeping, vegetable gardening), tourism and hospitality, entrepreneurs, community builders, sportsmen and women, pupils (best Grade 12 pupil in maths, science, IsiXhosa and agriculture) and healthcare workers (best clinic and best health worker). Sigcawu said that an award titled crimefree village will also be handed over to a deserved winner.

"Those who work hard need to be applauded so that others can be encouraged to do the same. This is more than a competition, it is to recognise commitment and dedication," he said.

Residents have applauded the initiative, saying it would definitely motivate them.

Female coordinator and [member of] Council of Stakeholders serving under the chiefs, Nolusapho Nqwane, said it was a pity that awards incentives had to be used to inspire a community. "Everybody should want to do their best for their community, knowing everyone will benefit." Nqwane, a school teacher, who also farms livestock said: "We have plenty of land from which to produce food, as well as the skills and talent for arts and culture. Our communities by nature are not lazy. But they have been discouraged by the lack of markets where they can sell their produce, unfenced fields which don't contain livestock, nor do they keep out those that would ruin their crops." However, she said: "We want our children to shine, both in education and at sport. I believe these awards will definitely tap hidden talent and inspire our children to do more."

Retired education development officer Toto Thetyana and senior Grahamstown police officer Dumisani Maphukatha, are just two of the many progressive farmers at Candu village. Maphukatha farms with 230 sheep and 40 lambs, 32 cattle, 120 goats, 10 pigs and more than 100 chickens. Thetyana, 69, who started farming in 1978 when he was a junior teacher, now owns 31 cattle, 255 sheep and 43 goats. Maphukatha and Thetyana between them plough more than 10ha of mielies. Thetyana also has about four tons of maize which he stores in grain silos to feed his livestock and both have their own tractors, with a full complement of farming implements.

These farmers and other entrepreneurs in the villages under Sigcawu, also employ other villagers—either as herders or tractor drivers. These people in turn are able to provide for their families.

Maphukatha, who is police colonel and head of support services in Grahamstown, and Thetyana each employ three shepherds and one tractor driver.

"Livestock is the treasure of any man. My father was an emerging farmer and he used the money he was making from farming to send us to school, right through to tertiary level. Cattle were sold to pay our fees," said Thetyana, who retired in 2001 as a school inspector. Maphukatha said he also earned money from selling wool. "Stock farming always gives you a 100% return, sometimes even 200%," said the policeman who bought his first lot of 10 sheep in 1992. Both Thetyana and Maphukatha said they were busy improving their livestock and over the past few years have been buying rams to improve their herds, including mutton Merino and Dohne merinos. "I have also bought Nguni bulls to improve my cattle," said Maphukatha, adding that his wife, a teacher, was managing the livestock in his absence.

Nokhaya Thetyana, born in 1939 at MaBheleni in Candu, is an emerging beef farmer, although at this stage only has nine cattle and one beautiful bull. "I bought one heifer and now have nine cattle. I could have more, about 20, if some had not died and others stolen. These awards are going to encourage the youth to move away from crime," said Nokhaya.

Sigcawu said he believed agriculture and education were key to rural development and needed to be encouraged and supported at all costs. "We must not have a nation that depends on government handouts. Our nation must have initiative and work with the little they have to do bigger things that benefit the community," he said.

Toto Thetyana, Maphukatha, Nokhaya and others have not received any grants from government. There are some people however, like the 10-member Gcaleka Youth Project, a brick-making entity, that started off six years ago with government funding.

Project chairman Bathandwa Madaka said: "As youth we did not want to sit and do nothing but we wanted to be involved in a project that would keep us busy. This shows you can do something on your own and succeed," said Madaka.

Despite the fact that bead-making is a skill that is mainly practised by women, Mncekeleli Mgoduka, 75, is a respected bead-maker. "I never got any training. I had a vision in a dream and was shown how to do beadwork. I have dressed many important people, including traditional leaders and politicians," said Mgoduka, better known by his clan name, Gcwanini. A father of 12, of Lotha Village, he is also a *sangoma* [traditional healer] and herbalist. He started beadwork in 1999 and has trained many people in the trade.

4.1. What Is in It for the Chief?

The first narrative we extract from this story is perhaps the most obvious. Why would a chief care about creating rewards for people living under his jurisdiction? After all, the chief's hold on office does not depend on citizens' behavior. He typically inherits the position of chief and, once attaining it, he retains it for life (e.g., Baldwin 2015; Mershon 2017; Williams 2010). Moreover, although analysts debate chiefs' capacity to enhance the provision of local public goods, they concur that chiefs' authority rests at least in part on coercion (e.g., Acemoglu et al. 2014; Baldwin 2013; Koter 2013; cf. Olson 1993). Afrobarometer survey data show that South African respondents hold more positive attitudes toward traditional leaders than toward elected officials (Logan 2009). In such a setting, why would a chief bother to set up incentives for citizen behavior of any kind? The first query is, then, ***Query 1***: With the specter of sanctions readily available, why does the chief institute rewards?[3]

To tackle this question, we pull the subset of the news story most relevant to it. In some ways, the *Daily Dispatch*'s depiction of an *nkosi* (chief), Xhanti Sigcawu, deepens the puzzle. Sigcawu is "impressed by his hardworking nation" and sees "talent and dedication," along with a "willingness to help others" among its members. Not surprisingly, Sigcawu proclaims that he is a "proud leader." As readers see in reading through Narrative 4.1, it is but a textual subset of the initial story. We limit ourselves to the facts in this newspaper story, maintaining consistency with our definition of narrative (chapter 2) as a subset of a story.

Narrative 4.1: What Is in It for the Chief?

Senior traditional leader and head of Mbhashe Traditional Council, Nkosi Xhanti Sigcawu, will reward those individuals [in] . . . his hard-working nation . . . who go the extra mile. . . . This gesture signals the beginning of an annual awards ceremony—the inaugural Mbhashe Traditional Council Awards.

. . . "We have decided to establish these accolades to honour and celebrate talent and dedication within our locality to motivate our people, so they can see the importance of commitment to development and excellence," said Sigcawu, who is the uncle of the AmaXhosa King, Mpendulo Sigcawu.

. . . Sigcawu, whose area of jurisdiction spreads west of the Mbhashe River to include the administration areas of Candu, Nywarha, Bholotwa,

Lotha, Nqabarha and Ludondolo, said his people's commitment, dedication and willingness to help others made him a proud leader.

"In most villages in other areas of the Eastern Cape land lies idle and the quantity of livestock has decreased. Even the livestock that is kept, is often of poor quality. However, my people are working the land, feeding themselves and others. The quality of livestock has improved tremendously over the years," . . . said Sigcawu.

. . . Sigcawu said he believed agriculture and education were key to rural development and needed to be encouraged and supported at all costs. "We must not have a nation that depends on government handouts. Our nation must have initiative and work with the little they have to do bigger things that benefit the community," he said. [Awardees] Toto Thetyana, Maphukatha, Nokhaya and others have not received any grants from government.

Given his expressed pride in his people as they are, why would senior traditional leader Sigcawu establish the reward system? Answers to this question lie in the chief's interactions with two sets of people: those who are under traditional authority, whom he rules; and those with some form of traditional authority, who may be senior traditional leaders, like him, or may be above or below him in the chiefly hierarchy.

4.1.1. The Chief as a Traditional Officeholder

First consider the people who do not wield traditional authority. *Nkosi* Sigcawu aims to cultivate the commitment of the people he rules. Although, again, he does not owe his office to their support, he depends on their loyalty, trust, respect, and deference. The chief's power rests on the primacy of traditional leadership for the members of his community. Sigcawu voices his own belief in and concern for the legitimacy and sovereignty of traditional governance structures by repeatedly referring to the people he rules as a nation or even "our nation." Traditional governance is an institutional structure distinct from electoral authority as it binds together a system of power and a set of practices passed down across generations, and is, presumably, sovereign over everyone in a traditionally defined nation. Moreover, while Sigcawu might care about coproducing their collective identity with community members as a matter of public interest, we assume that any such interest is consistent with his self-interest: the more his people see themselves as a nation, the more his authority is enhanced (Baldwin 2015; cf. Olson 1993, 2000; Ostrom 1996). In addition, the tangible collective identity is a sort of public good for other traditional officeholders, and thus buttresses the unity and comity among the leaders beneath him in the traditional hierarchy, the chiefs, headmen, and headwomen in his area of jurisdiction.[4]

Nkosi Sigcawu also values the demonstrated backing of those he rules because it strengthens his hand in relationships with those at his level or above him in the traditional hierarchy. In particular, the economic advances registered within his jurisdiction as compared to other jurisdictions distinguish him horizontally from other senior traditional leaders, and they positively recommend him to his superior, the AmaXhosa king, Mpendulo Sigcawu. As *Nkosi* Sigcawu is a close elder relative—an uncle—of the current king, the economic accomplishments of the community he names as "our nation" and "my people" potentially put him in a position of leadership among senior traditional leaders in his nephew's kingdom. In this light, his institution of new forms and procedures such as the awards ceremony may be seen as exercising and embodying leadership by example.

Viewing the same set of relationships in reverse, a senior traditional leader whose followers seemed reluctant would be weakened in any effort to exert his capacity to rule within his community and to display his authority to the chiefs and headmen and headwomen below him. How can he wield influence over traditional leaders he ostensibly outranks if he cannot show that his people approve of him and put their approval into action by producing public goods for his and their community? The awards ceremony puts in high relief the accomplishments of his subjects, indicates their investment in the community under the chief, and, by extension, suggests popular support for traditional institutions. In carrying out the awards ceremony, the chief institutes a new practice that can be emulated by other traditional leaders both above and below him, and thus establishes himself as an institutional innovator and creative statesman.

In addition, *Nkosi* Sigcawu cares about making clear to the king above him that he enjoys support from his people and authority over them. This concern could be all the more pressing—and delicate—given that he is the uncle of the king directly above him in the traditional hierarchy, the AmaXhosa king, Mpendulo Sigcawu.

We also need to ask why *Nkosi* Sigcawu paints government "handouts" as undesirable. To judge from his explicit affirmations, Sigcawu does not like it when his subjects turn to elected authorities for help. He states that the "handouts" threaten his community's independence, initiative, and capacity to benefit the collectivity. It is reasonable to extrapolate from these explicit statements and assume that the chief sees "handouts" from elected officials as threats to his right to rule. He does not want his subjects to become economically beholden to the elected government, for that would mean that they would shift the primary focus of their loyalty and trust away from him and to the elected government. By instituting the annual awards ceremony, *Nkosi* Sigcawu seeks to bolster—even more, boost—his authority in the face of elected officials and bureaucrats dependent on elected government. The label he uses of "government handouts" suggests that receiving them is a sign of personal weakness.

What is more, the newly instituted award ceremony is supposed to extend into perpetuity, signifying its superior durability relative to any policy initiatives of elective governments. The chiefs' time horizon exceeds that of elected politicians: they rule as chiefs for life. Their long tenure strengthens their ability and incentive to facilitate the cooperation among citizens underpinning the supply of public goods. Chiefs should thus mobilize action among citizens under their jurisdiction to yield shared local benefits. Chiefs with citizens should, in a word, "co-produce" public goods (Baldwin 2015; cf. Ostrom 1996).

4.1.2. Model 4.1: The Chief's Utility Function

Several features have emerged from the first narrative that can be "translated" into components of the utility function of the chief. While working with the story, we discovered his dislike of his community's reliance on "handouts" from the elected government. We choose to interpret that observation as the chief's preference for raising the importance of traditional vis-à-vis elected government in the lives of the members of his community. To capture this preference with notation, label as χ, *s.t.*, $\chi \in [0, 1]$, the weight in the overall perceived legitimacy that the community members place on the traditional structures, and thus as $1-\chi$ the weight that they place on elected governments. What we have just stated is that the utility of the chief increases in parameter χ, or $\partial u/\partial \chi > 0$. The parameter χ appears on the first row of table 4.1, which summarizes all the notation used in this chapter.

Our next observation was that Chief Sigcawu proclaimed the superiority of his nation's economic achievements relative to those of the other subjects under his nephew, the King. If true, then Chief Sigcawu deserves praise and recognition above and beyond that going to the other chiefs. The Chief wants such praise and recognition, as it lifts him above his peers within the traditional hierarchy: keep in mind the way intermediate steps in the hierarchy are defined in his world, in terms of seniority, wisdom, sagacity, and entitlement to impart knowledge and instruct others on action. We can call that aspiration vertical advancement, and denote it with the notation v. This parameter positively affects the Chief's utility, so $\partial u/\partial v > 0$ as well.

Not only is the Chief rewarding his trailblazers, but he is also a trail-blazer himself, in terms of pioneering a new institutional form of annual awards with accompanying broad recognition. We reason that in doing so

TABLE 4.1. Table of notations

Symbol	Meaning
$\chi \in [0,1]$	weight of the perceived legitimacy that community members place on traditional structures
v	aspiration for vertical advancement on the part of the Chief
h	aspiration for horizontal influence on the part of the Chief
p	(non-zero) probability of the collapse of the Kingdom ruled by the Chief's family in each period (e.g., each decade)
d	durability of traditional institutional forms as perceived by the community
m	promise of a monetary award to be given to some community member in the future
u_C	utility function of the Chief
W_a	utility function of villagers who receive awards
s	a villager's upward social move (standing)
z	a villager's enhanced horizontal visibility among other villagers
δ	discount factor for the stream of future benefits from the award ceremony, i.e., the survival probability of the tradition of that ceremony
W_0	utility function of villagers who have not received awards
g_s	subsistence agricultural investment
g_m	market agricultural investment
g	total agricultural investment made by a villager
$\alpha \in [0,1]$	weight placed on subsistence agriculture in a villager's farming allocation
$\gamma_s \in [0,1]$	probability of realizing g_s
$\gamma_m \in [0,1]$	probability of realizing g_m
$\gamma_{m'}$	updated probability of success of market investment
i	income from a salaried position in the community
$\varphi \in [0,1]$	probability of attaining i
e	impact of education on becoming a salaried employee
c	crime rate

Note: Notation appears in the order it is used in the chapter.

he seizes the role of an innovator, a leader, within his hierarchy, and, most importantly, relative to his peers—other chiefs. Chief Sigcawu would like to be followed and emulated throughout the kingdom, so that other traditional leaders would look up to him for example and guidance. Calling this goal that of horizontal influence, we will label it h, another parameter in which the Chief's utility is increasing ($\partial u/\partial h > 0$).

Finally, because we discovered the Chief's very close familial relationship to the King, we can theorize that the Chief's stakes in the long-term preservation of the kingdom as a unit and form of government in general and of his nephew's lineage in power in that kingdom in particular are very important concerns. How can we capture this twofold consideration in the Chief's utility function? Suppose there is a nonzero probability of the collapse of his family's regime in each period; suppose that a period is a decade. Call this probability $p > 0$, and observe that anything that diminishes this probability is therefore good for the resilience of the regime. Some of the chief's actions may contribute to reducing p. Specifically, he has just put in place a new institutional form that will presumably last indefinitely and will not be subject to reversions of policy decisions enacted by elected governments (d, for the durability of institutional forms as perceived by the community). The promise of a monetary prize to go with the award in the future (m, for money) may also contribute to the King's subjects' happiness with the rule and so with the ruler. What we now have is the probability of regime collapse declining in such variables, $\partial p/\partial d < 0$, and $\partial p/\partial m < 0$. Since we also know that the chief fears the regime's collapse, his utility in turn declines in p, $\partial u/\partial p < 0$, which implies that it increases in d and m.

To summarize, we have specified the utility function of the Chief as

$$u_C(\chi, v, h, p(d, m)) \tag{4.1}$$

To reinforce the connections running throughout the book, observe that in reaching expression 4.1 we have traveled for the first time along the N → M route indicated in figure 2.1, from chapter 2. Moreover, expression 4.1 features the primitives of the utility maximization approach as outlined in table 3.1: a single actor operates, she faces uncertainty, her utility is assumed to be cardinal, and her beliefs are described as probability weights on states of the world. We do not, however, fully specify a constrained maximization problem—as we would do if we were to choose to develop a decision-theoretic model. Instead, we proceed with a general discussion of some comparative statics that give immediate rise to statistical hypotheses. All models in this chapter share these features, as we will see.

4.1.3. A Few Testable Implications from Model 4.1

The model highlights several determinants of chiefs' behavior that also appear in the relevant literature. Even though the literature dwells heavily on the capacity for coercion as a source of chiefly authority, scholars also acknowledge that chiefs depend on some degree of deference, loyalty, and trust from the people they rule (e.g., Baldwin 2015; Koter 2013). Next, chiefs want to increase the importance of traditional vis-à-vis elected government for community members, and fear reversions of traditional policies at the hands of elective authority (cf. de Kadt and Larreguy 2018; Mershon and Shvetsova 2018, 2019; Williams 2009, 2010). Additionally, they aim for vertical advancement and horizontal influence within the traditional hierarchy (e.g., Baldwin 2015; Williams 2010). There may also be a concern, as in the case in the story, for the survival of a specific regime at the helm of the kingdom. This means that chiefs might care about their communities' perceptions of an incumbent king's regime and the benefits associated with it.

We now move to sketch testable implications for some hypothetical universe of chiefs and their communities. The hypotheses as we illustrate them do not exhaust the possibilities, but rather drive home the insights to be gained by applying the algorithm for narratives and models.

H4.1.1: Chiefs should endorse any move from elected officials to entrench traditional authority.

H4.1.2: In areas with strong chiefly authority, individual villagers should receive fewer government grants than in other areas, other things equal.

H4.1.3: Chiefs in the immediate family of the king should engage in more institutional innovation than chiefs outside the immediate family, other things equal.

4.2. What Is in It for the Villagers?

The second narrative shifts the focus to the community members under the chief's jurisdiction. According to the chief, villagers "need to be applauded." But that claim begs the question: Why? Why would villagers be induced by awards to contribute to their own community? Are villagers not motivated by their concern for their family, their view of their "nation" as a larger variant of their extended family, and their sense of shared fate with this community? These questions culminate in *Query 2*: Why do the fruits of contributing to the common good not suffice as incentives to the villagers? Consider this query in reading Narrative 4.2.[5]

Narrative 4.2: Why Would Villagers Be Motivated by Awards?

Senior traditional leader and head of Mbhashe Traditional Council, Nkosi Xhanti Sigcawu, will reward those individuals who go the extra mile. "We have decided to establish these accolades to honour and celebrate talent and dedication within our locality to motivate our people. . . . This is more than a competition, it is to recognise commitment and dedication," he said. Residents have applauded the initiative, saying it would definitely motivate them.

Female coordinator and [member of the] Council of Stakeholders, Nolusapho Nqwane, . . . a schoolteacher, . . . said . . . "We want our children to shine, both in education and at sport. I believe these awards will definitely tap hidden talent and inspire our children to do more."

Retired education development officer Toto Thetyana and senior Grahamstown police officer Dumisani Maphukatha, are just two of the many progressive farmers at Candu village. Maphukatha farms with 230 sheep and 40 lambs, 32 cattle, 120 goats, 10 pigs and more than 100 chickens. Thetyana, 69, who started farming in 1978 when he was a junior teacher, now owns 31 cattle, 255 sheep and 43 goats. Maphukatha and Thetyana between them plough more than 10ha of mielies. Thetyana also has about four tons of maize which he stores in grain silos to feed his livestock and both have their own tractors, with a full complement of farming implements.

. . . Maphukatha, who is police colonel and head of support services in Grahamstown, and Thetyana each employ three shepherds and one tractor driver. . . . Maphukatha said he also earned money from selling wool. "Stock farming always gives you a 100% return, sometimes even 200%," said the policeman who bought his first lot of 10 sheep in 1992. Both Thetyana and Maphukatha said they were busy improving their livestock and over the past few years have been buying rams to improve their herds, including mutton Merino and Dohne merinos. "I have also bought Nguni bulls to improve my cattle," said Maphukatha.

. . . Mncekeleli Mgoduka, 75, is a respected bead-maker. "I never got any training. I had a vision in a dream and was shown how to do beadwork. I have dressed many important people, including traditional leaders and politicians," said Mgoduka, better known by his clan name, Gcwanini. A father of 12, of Lotha Village, he is also a sangoma [traditional healer] and herbalist. He started beadwork in 1999 and has trained many people in the trade.

Consider, too, that the second narrative shifts the time horizon. Why would awards have an immediate impact on the villagers receiving them? Such economic activities as beef farming involve uncertainties and difficulties, as the record of dead and stolen cattle makes clear. What can the visibility of an award do to overcome such obstacles? Or might the recognition in the form of the "accolade" affect the villagers' willingness to confront the obstacles entailed in farming?

4.2.1. The Villagers' Responses to the Awards

The first step in understanding the villagers' responses to the awards is to acknowledge that there are two types of villagers: those who earned an award at this first annual ceremony and everyone else. The benefits to these two groups in the community will differ in their sources.

The awardees. The moment of the award ceremony brings public recognition to the recipients, who can be proud of their accomplishments. These individuals can expound at length on their knowledge of, and skill in using, new agricultural methods. They now act as teachers, instructing their neighbors on the right path to prosperity. In a traditional society, experience in teaching the rest means advancing in social standing. Even after the occasion itself, the award recipients can enjoy their enhanced standing in the community. In addition, with another round of awards bestowed each year, at another ceremony, the award recipients from any given year gain indirect recognition for their past accomplishments, as the new ceremony triggers their own and others' memories of the prior ones.

The individuals in the story are nodules in community networks, and the awards strengthen their position as key nodules. Some people both hold valuable salaried positions such as police officers and simultaneously provide for the community through their entrepreneurship in stock farming. This means protein for neighbors and enhancing community health, as well as employment and greater engagement with the market economy. Others, like the bead-maker, Mgoduka, who has "dressed many important people," preserve the community's customs and cultural distinctiveness. The beads he weaves into traditional dress reinforce the aura of power of the hereditary elites wearing the garb and also reaffirm the shared identity of the traditional community. The nod to Mgoduka is important. He personifies mastery of the traditional trades not only through bead-making but also as a traditional healer and an herbalist. Mgoduka is known by his clan name, suggesting his role as a respected elder and a fixture in the community.

The award ceremony broadcasts to the community the signal achievements of particular individuals and gives them the right to be proud, brag, and maybe even teach others how to develop the abilities celebrated in the award. The ceremony allows award recipients to publicize in detail their livestock owned, herds improved, hectares farmed, profits earned, and children enrolled in school. Spreading the news of their talents and achievements serves to increase their standing in the community. For other awardees, such as Mgoduka, the ceremony bolsters the esteem they enjoy in the community, along with their position as a trusted fount of knowledge, experience, and wisdom. All awardees, and the chief, can hope that people in the community will follow their good examples.

Additionally, if and for as long as this current chief remains at the helm, the awards will continue, and the ceremony will gain even more prominence, so that the awardees will receive a stream of future benefits. Their standing as "elders" who are wise and who teach, and their fifteen minutes of fame, will get refreshed like clockwork, every year, either by simple association with the current awardees, or even from sitting on the stage during future iterations as past recipients and living evidence of the value of honors bestowed on those next groups.

The nonawardees. All villagers must meet basic life needs in some way. Villagers want if at all possible multiple income streams to diversify and lower the economic risks they face. Subsistence agriculture ties villagers to their homes but carries relatively high risk. They thus want marketable agricultural products with a relatively high and stable return, for example, such improved breeds as Maphukatha's sheep.

Villagers who did not receive awards still receive valuable information about what might be labeled real-world experimental trials by their more risk-accepting peers, the awardees. The awards ceremony not only advertised successful experiments but also invited bragging about the details—the technology that enabled those accomplishments. This is how they convey information about risks and benefits—that is, about the rewards of investing in particular technologies, and how those technologies also reduce market risks. With this information, the nonawardees can update their beliefs about the value of branching out into market agriculture. Note that having trailblazers who were able to act as early adopters makes it easier for the rest of the villagers to take risks and to diversify their income sources (or, equivalently, diversify their assets). If the nonawardees follow the example of the awardees, their risks can be considered acceptable; one of the award recipients goes so far as to claim that there are no risks.

Having observed the experience of the awardees, the nonawardees now consider a lower probability of failure and thus a higher expected return from market agriculture. At the same time, the probability of failure in subsistence agriculture remains unchanged. Therefore, it benefits the villagers to "rebalance their assets," moving in part or in full from subsistence to market-oriented agricultural production. If and when they do that, as they now can, given their new information, they are going to be both wealthier and safer. In sum, the villagers not winning an award perceive lower overall risk and higher overall expected return on their investment, broadly conceived. They benefit by learning from their peer role models, the awardees.

Note that all the "trailblazers" in market agriculture that the story mentions are also salaried employees, and this income security may be what has allowed them to take risks and get ahead. The villagers know that education is necessary for holding such salaried jobs as school inspector and police colonel. This reinforces the villagers' understanding that a good life requires attention to school. They want their children to have the option of complementing livelihoods on the land with some other, more secure and steady source of income. The villagers want their children to flourish, and they link that with educational attainment. The recognition of excellence in academics among schoolchildren, then, clearly serves the villagers as mothers, fathers, aunts, uncles, and grandparents.

4.2.2. Model 4.2: The Utility of Two Types of Villagers

One element of the utility function of award recipients, W_a, comes from their bragging rights and their new role as informal teachers; it is their upward social move, standing, s, which rises with receiving an award and increases their utility, $\partial W_a/\partial s > 0$. Another element is their enhanced visibility, z, which distinguishes them horizontally from their peers, that is, makes them famous in the community. We are going to assume that they rather like more visibility: $\partial W_a/\partial z > 0$. The award recipients also care about the stream of future benefits, discounted by some factor δ. The discount factor can be interpreted as the survival probability of the tradition of this ceremony, and will itself depend on the stability of the current ruling traditional regime, labeled above as p. The awardees benefit in the long term from a higher discount factor, so $\partial W_a/\partial \delta(p) > 0$ and $\partial W_a/\partial p < 0$. We thus specify the award recipients' utility function as:

$$W_a = \sum_{t=0}^{n} \delta^t(p) W_a^t(s,z) \qquad (4.2.1)$$

The utility function of those villagers who have not earned awards we can denote as W_0. As conjectured above, all villagers want to be in secure and comfortable material circumstances. Since subsistence agriculture, g_s, in which they all are traditionally involved, is a risky endeavor, they seek, if possible, to have multiple income streams to diversify and lower the economic risks they face. Supposing that their total agricultural investment is fixed (e.g., by size of land holding), one consideration for them is whether to diversify into market agriculture, $g_m = (1-\alpha)g$, such that $\alpha g + (1-\alpha)g = g$, where g is the total agricultural investment of a villager, and $\alpha \in [0, 1]$ is the weight that the villager places on subsistence agriculture in his or her farming allocation. The way a farmer would allocate agricultural investment would depend on the expected returns, that is, on the size of output and the probability of successfully obtaining that output from each type of investment. Consider interpreting g_s and g_m, then, as the profitability of subsistence and market agricultural investment, respectively, and denote as γ_s and γ_m the respective probabilities of realizing such profit, $\gamma_s, \gamma_m \in [0, 1]$.

While all villagers are familiar with the probability of crop failure in subsistence farming, $1-\gamma_s$, from their extensive personal experiences, they evidently consider the probability of market agriculture failures to be very high, $1-\gamma_m \gg 1-\gamma_s \Rightarrow \gamma_m \ll \gamma_s$, because we can see from the story that engagement in market agriculture is relatively rare among them.[6] When the awardees brag and so disseminate their knowledge of market agriculture, the rest of the villagers can update their beliefs about its riskiness: $\gamma_m < \gamma'_m$, where γ'_m is the updated probability of success of a market investment. If the expectation of success is thus sufficiently improved for the farmer to embark on market farming, then he or she would do so, and the farmer's utility would increase: $W_0(\gamma'_m) > W_0(\gamma_m)$. Generally speaking, W_0 is nondecreasing in γ'_m.

A way to diversify and add to a villager's stream of income is to achieve a salaried position in the community. Denote the income from such a position as i, and the probability of holding such a position as $\varphi \in [0, 1]$. Then, insofar as the award ceremony helps motivate students in their educational efforts and supposing that education, e, improves one's chances of becoming a salaried employee, villagers' expected utilities increase with educational attainment, $\partial W_0/\partial e > 0$.

Another element in all villagers' utility function is their utility from the public goods provided within the community. Crime prevention is one such public good, and the award ceremony, which encourages youth to focus on schoolwork and sports, is expected to lower the crime rate, c, and so increase the utility of an average villager.

Altogether, the utility function of a nonawardee villager can be expressed as

$$W_o(\gamma'_m, e, c) ,\qquad\qquad (4.2.2)$$

where it is nondecreasing in γ'_m, increasing in e, and decreasing in c. The impact of conducting the award ceremony on all its determinants is such that the utility of a nonawardee goes up as a result.

4.2.3. Testable Implications from Model 4.2

As it underscores inducements to the choices made by members of communities under chiefly jurisdiction, Model 4.2 also aligns with extant relevant research (e.g., Baldwin 2015; Beall 2005, 2006; Ensminger 1996; Oomen 2005). The story and the model make clear that "trailblazers" are risk takers. The diversification of income streams makes prominent community members better able to experiment in agriculture, and their achievements can then be disseminated in the community. We can thus safely conjecture that villagers recognized by chiefs should be relatively likely to expand their local and extralocal networks. It is also reasonable to conjecture that villagers, whether or not they gain chiefly recognition, should benefit nonetheless. Such benefit could materialize, for instance, by their increased confidence about entering market agriculture or encouraging their children to prepare for salaried nonagricultural employment.

In this way, the model leads to testable implications for a hypothetical universe of communities ruled by chiefs as well as a larger separate set of all communities, both ruled by chiefs and not. Once more, we intend for the illustrative hypotheses to convey the analytic leverage to be had by adopting the algorithm for narratives and models.

H4.2.1: Villagers who receive awards should be more likely than other villagers to expand their local and extralocal networks in the next period, other things equal.

H4.2.2: Communities where chiefs generate individual recognition for market-economy pioneers should see higher rates of economic development in the next period than do other communities, other things equal.

4.3. Does Turning to Awards Point to an Erosion of Traditional Identity?

Our third narrative brings together the chief and the villagers. We remain interested in why the chief believes he must institute the annual awards ceremony so as to "recognise commitment and dedication." We still ask, too, why the villagers require motivation from the rewards. But does the chief's use of a reward system point to an erosion of the villagers' identity as members of a traditional community, a community defined by shared values, norms, and customs, ruled by traditional authority, and distinguished as a nation within its boundaries? In sum, we pose *Query 3*: How much does the establishment of an annual ceremony amount to a defensive measure designed to fortify an otherwise fading collective identity? Keep in mind this query in reading Narrative 4.3.

Narrative 4.3: How Much Do Awards Signal an Erosion of Traditional Identity?

Senior traditional leader and head of Mbhashe Traditional Council, Nkosi Xhanti Sigcawu, will reward those individuals who go the extra mile. Thirty-six individuals . . . will be rewarded with a trophy and have their names inscribed on an honours board for posterity at a special ceremony today. This gesture signals the beginning of an annual awards ceremony— the inaugural Mbhashe Traditional Council Awards—at the Mbhashe Royal Village Lodge on the banks of the Mbhashe River.

. . . Sigcawu . . . said his people's commitment, dedication and willingness to help others made him a proud leader. "In most villages in other areas of the Eastern Cape land lies idle and the quantity of livestock has decreased. Even the livestock that is kept, is often of poor quality. However, my people are working the land, feeding themselves and others. The quality of livestock has improved tremendously over the years . . . ," said Sigcawu. . . . An award titled crimefree village will also be handed over to a deserved winner. "Those who work hard need to be applauded so that others can be encouraged to do the same. This is more than a competition, it is to recognise commitment and dedication," he said.

. . . Nqwane . . . [female member of the] Council of Stakeholders . . . , said it was a pity that awards incentives had to be used to inspire a community. "Everybody should want to do their best for their community, knowing everyone will benefit. . . . We want our children to shine, both in education and at sport. I believe these awards will definitely tap hidden talent and inspire our children to do more."

... Nokhaya Thetyana ... is an emerging beef farmer, although at this stage only has nine cattle and one beautiful bull. "I bought one heifer and now have nine cattle. I could have more, about 20, if some had not died and others stolen. These awards are going to encourage the youth to move away from crime," said Nokhaya.

We start by highlighting the venue of the awards ceremony, the Mbhashe Royal Village Lodge, by the Mbhashe River. The ceremony gathers Sigcawu's nation in a location that showcases their common heritage. We can infer, though the story does not explicitly state, that the Mbhashe Royal Village Lodge provides the natural place for displaying the "honours board for posterity." The location says that the awards initiative is not a personal project of Sigcawu but belongs to the Mbhashe Traditional Council and the people in the community more broadly. The venue communicates, too, the chief's and council's authority in the community. To judge from the story, the villagers when discussing the awards see the preservation of traditional identity as valuable above all for children and youth, for the next generation.

4.3.1. Redefining a Contemporary Traditional Community

Consider the elements of the narrative that suggest that, on the one hand, traditional identity might be weakening, and, on the other, that awards might bolster traditional identity. Such concerns and hopes are two sides of the same coin. Observe first how *Nkosi* Sigcawu boosts the morale of his people by praising their superior economic achievements when contrasted with "most villages in other areas" of the province.

In the way that he words his praise, Sigcawu attributes the success to the entire nation, sending the message that his nation is both vital and vibrant. Nonetheless, all awards, except for a few explicitly devoted to community building, are given to flag individual achievements and individual prosperity, which are thus viewed as contributions to the community. In so designing the reward system, the chief moves to reconcile new market economy individualism with traditional collectivism in terms of both goals and achievements.

If villagers associate the upholding of traditional values with personal costs when it comes to economic choices, and if this association makes them discount the entire nexus of traditional values and beliefs once they stop conforming with a few of those through their economic behavior, then the Chief's strategy may be effective. By publicly declaring what used to be seen as economic nonconformism as instead now legitimate community behavior, he removes the all-or-nothing choice from the hard-pressed villagers.

The theoretical angle in the construction of this last narrative is that the preservation of traditional community values is a public or collective good, and its provision should require collective action. Collective action, in turn, carries with it a host of associated problems of free riding, as well as conflict over division of effort among the members of the group (cf. Hardin 1971, 1982; Olson 1965). With expanding opportunities in the public sector and market agriculture, do we now witness a clash between the needs of individuals and families, on the one hand, and, on the other, the expectations of the traditional leader and of the community that espouses traditional values?

An elderly "emerging farmer," Nokhaya Thetyana, brings up encouragement from the awards. But he refers to crime as the reason that encouragement is needed. He reports stolen cattle, which indicates the erosion of such traditional values as trust and honor. Since he anticipates that the awards will lead youth away from crime, we must assume that he has in mind the awards to schoolchildren in education and also in sports. More broadly, awards for young people's achievement in school and sports promise to foster their prosocial behavior. Echoing this concern, Sigcawu establishes a specific prize named crimefree village.

We can also interpret the Chief's concern about reliance on government "handouts" as broadly implying a fear of free riding, of not making enough of an effort on one's own. This worry about free riding, along with anticipations of the awards' influence, suggests that the villagers may vary among themselves in the weight they place on traditional values and norms.

Sigcawu and the villagers have a shared stake in education and youth. He underscores its importance for advancing rural development. Awards to schoolchildren for their excellence in education not only shape the next generation but also benefit everyone. We are ready to suppose that a better educated next generation is a separate component in the nation's system of values. The continuation of the awards into the future, in an ongoing annual ceremony, is especially important as a motivation for young people and thus pleases those who parent and guide them.

The Chief seems to be aiming to dispel the perception of tradition as standing in the way of progress and of individual advancement. This stance of his possibly contrasts with what his more conservative constituents may believe, as he pushes an interpretation of individual advancement as community service. Indeed, he prioritizes steps toward economic prosperity as the most valuable contribution to the relative standing of his nation. Villagers like the bead-maker, Mgoduka, would naturally form part of the Chief's conservative constituency. Placating such constituents with awards lowers the cost to the chief of introducing his new vision.

How to manage the health of traditional values in the context of the new economy and the changing society? Can this be done without diminishing individual entrepreneurship and the ambition to advance? It looks like the awards carry a host of incentives to various types of community members, and so may readjust behaviors and attitudes in the desired direction. Nqwane claims that it is "a pity" that awards must be used. She implies that some villagers should do more for the community out of a sense of duty; otherwise, they ride free on the efforts of others. Chief Sigcawu seems to take a more realistic and pragmatic approach. He balances embracing individualism as the driver of economic development and prosocial behavior and an example for emulation, on the one hand, with, on the other, the incentives and positive reinforcement he adds for observing and preserving traditional values, practices, and skills.

In line with the need to placate his conservative constituency, the Chief chooses to recognize schoolchildren's mastery of IsiXhosa, which embodies and perpetuates the community's traditions and continues to bind the villagers together and to their leaders. Even the grade level is meaningful: awards go to students in grade 12, the last year of secondary school. This hints at a possibility that the goal of the prize is to bring up the next generation of prospective village leaders steeped in custom and tradition, to identify the youth with the strongest interest and deepest investment in the nation.

4.3.2. Model 4.3: Villagers' Compliance with (Redefined) Community Values

In this model, the premise is that the chief prefers to reign over an enthusiastically compliant nation rather than a nation of free riders, and if attaining that goal takes redefining the notion of what the tribal public good is, then so be it. We can borrow, for convenience, the utility function of the Chief as we designed it in Model 4.1 to illustrate how the specific content of the tribal good may be of little importance to him, compared to the instrumental benefits from having happy and compliant constituents. The function that we used there was $u_C(\chi, v, h, p(d, m))$, where χ, v, h, p, d, and m stand for, respectively, the weight in the overall perceived legitimacy that the community members place on traditional structures, the Chief's vertical advancement due to his greater institutional "activism" than that of his peers, his horizontal influence stemming from his support within, and the wealth of, his nation, and the survival probability of the dynasty in power to which he belongs. Observe that then, as well as now, we see no need to include the deeply ingrained tribal values anywhere on the Chief's list of priorities. Thus, in order to maximize his utility function as we specified it, the Chief should be willing to sacrifice traditional tribal values, redefining what they mean in contemporary practice.

The original conception of the tribal public good may have become nonimplementable, for a variety of reasons. We could suppose that the coercive ability of traditional governance has declined, which means that enforcement is now lacking. Alternatively, we could make an assumption that individual opportunity costs to villagers have increased with the arrival of new economic advantages to pursue. In either case, the implementation of old mechanisms for the provision of the tribal good has degraded: the mechanism has become nonimplementable for a significant portion of the agents (the villagers). Supposedly, when a chief redefines the notion of the tribal public good, he redesigns the mechanism to make it meet the implementation constraint: to make it such that an individual agent would prefer to follow tribal requirements rather than violate them, considering the ensuing punishments and rewards she foresees (e.g., Acemoglu et al. 2014; Koter 2013).

The way to model this logic would be to compare the implementation constraint of a villager before and after the chief redefines what it means to contribute to the tribal public good. Specifically, a villager's implementation constraint is expressed as

$$u(c(e), m, \chi, h, p(d, m)) \geq u(c(\neg e)) \tag{4.3}$$

which incorporates elements of expressions (4.1), (4.2.1), and (4.2.2). Thus, the villager derives benefit from the institution of the awards ceremony, as treated in (4.1), with its effect on the relative weight of traditional structures in the community's perceptions (χ), the Chief's horizontal influence (h), the probability of the collapse of the traditional regime (p) as a function of the ceremony's durability and its promise of future monetary prizes (d, m). The probability of the regime's collapse affects the utility of a villager who receives an award, expression (4.2.1); the utility of a villager who receives no award is a function of the crime rate she experiences and the education she has ($c(e)$), expression (4.2.2).

4.3.3. Testable Implications from Model 4.3

The third model, too, accords with several recurring themes in current scholarship. The issue of erosion of traditional values is conceptualized at least implicitly in terms of collective action. As suggested, the predominant perspective in the scholarship on traditional authority is that chiefs influence, monitor, and (threaten to) sanction citizens (cf. Acemoglu et al. 2014; Koter 2013; Mares and Young 2016). Acknowledging chiefly coercion yet advancing a distinct view, Baldwin (2015, 21, 10) characterizes chiefs as "socially and economically embedded [local] leaders" and "development brokers." The chiefs' long tenure strengthens their ability and incentive to facilitate the cooperation among citizens underpinning the supply of public goods. Strong chiefs should thus take initiatives that yield benefits to their subjects at the community and individual levels (cf. Baldwin 2015; Olson 2000; Ostrom 1996). Our conjecture is that chiefs, embedded as they are in their communities, act to buttress villagers' individual initiative by embracing that individual initiative as service to the community. In so doing, chiefs remove disincentives to participation that may come from calls for personal sacrifice for the sake of the community. Chiefs thus redefine and renew traditional communities in the contemporary era.

The following additional hypotheses illustrate the analytic leverage we have now gained.

H4.3.1: Where chiefs innovate institutionally, aggregate economic development (or prosperity) should be relatively high (great), other things equal.

H4.3.2: Where chiefs innovate institutionally, secondary schoolchildren should attain higher scores on standardized educational tests than in other areas, other things equal.

H4.3.3: Where chiefs innovate institutionally, surveys should reveal greater legitimacy of traditional authority than in other areas, other things equal.

H4.3.4: Where chiefs innovate institutionally, measures of perceived conflict between individual well-being and traditional values should be lower than in other areas, other things equal.

The reader would probably think of a few more hypotheses at this point, but we stop here.

4.4. Conclusions

The models designed in this chapter are simple in the extreme: we have repeatedly heeded Occam's instruction to razor away complexity. Rather than viewing the plainness as problematic, we see it as advantageous. Even models as spare as those in this chapter test theory. Each has made explicit all of our assumptions about each parameter we have brought to the forefront, in terms of its impact on the actor's utility. In doing so, each model has enabled us to examine whether any assumption is superfluous or inconsistent with any other. We also find from the models in chapter 4 that modeling enhances hypotheses. For instance, we are aware of no hypothesis paralleling H4.3.4 in the extant available literature on chieftaincy, whatever the methodology used.

As chapter 3 has demonstrated, types of formal models differ in crucial ways, making some types complex and others simple. The basic utility maximization approach, as we have observed here, is quite spare. This approach means making assumptions about aspects of actor preferences that drive their behavior. Chapter 5 illustrates the use of decision theory as we take a story of regulation of fishing and fashion several models on the conflict between individual interests and the common good.

FIVE

Industry Regulation

Rationale and Stakeholders

With Julie Vandusky-Allen, *Boise State University*

In this chapter, as in the last, we begin with one news story and next extract multiple narratives based on alternative interpretations of the story. Each narrative in turn supplies the material for a model.[1] Unlike chapter 4, where we created utility maximization models, here we design models that are decision-theoretic. As we indicated in chapter 3, decision-theoretic means that the actor makes her most advantageous choice out of a well-defined set of feasible choices. This mix of similarities and differences is not incidental but rather is intentional. We again implement the algorithm for narratives and models. We again evaluate and adjust the fit between, on the one hand, the modeled universe of logical steps and, on the other, the theoretical universe of premises and conjectures.

Alaska's most widely read newspaper published the story about fishery management that provides the basis for this chapter. The *Alaska Dispatch* covered the interactions among commercial fisheries, tribal groups, U.S. federal agencies, state actors, and regional regulators as they considered whether to place an emergency cap on the accidental catch (or bycatch) of Chinook salmon in the Bering Sea. As the story conveyed, residents in communities all along Alaska's two principal rivers had a stake in whatever decision would be reached, as did commercial fishers. The governmental and regulatory bodies involved faced a daunting challenge, given the clash of interests, the reliance on different sets of information by different actors, and the multiple principles that could be used in adjudicating among interests. With so many actors and so much conflict, this story clearly fulfills our selection criteria.

Story:

Tribal Groups Seek Emergency Action to Cap Accidental King Catch

Lisa Demer

Alaska Dispatch News | 2014-09-17

Two leading Alaska Native tribal organizations on Wednesday petitioned the federal government to dramatically lower the cap on the number of king salmon that Bering Sea commercial fishermen can harvest as bycatch in order to protect the fish.

The Association of Village Council Presidents and the Tanana Chiefs Conference filed their petition with the U.S. Department of Commerce secretary and the North Pacific Fishery Management Council for an emergency cap they say is needed to avoid substantial harm to the kings, or chinook salmon, and to communities up and down the Kuskokwim and Yukon rivers, the two biggest in Alaska.

The tribal groups want the government to lower the hard cap on the accidental catch of kings during the lucrative Bering Sea pollock fishery from 60,000 to 20,000 for the rest of 2014. There's a lower bycatch number that triggers tighter monitoring, and the groups also want to see it dropped, from 47,591 to 15,000.

A "multi-year downward spiral" in king populations "present(s) a serious conservation and management problem requiring immediate emergency measures on the high seas as well as the river systems," the 11-page petition said.

This year, directed subsistence fisheries for kings were completely closed on both the Kuskokwim and Yukon rivers.

If village residents who depend on salmon for survival can't catch kings, commercial fishing interests shouldn't be able to either, said Myron Naneng, president of the Association of Village Council Presidents.

A 20,000-king cap would only have been exceeded once in the last five years, so putting it in place won't unfairly restrict the Bering Sea fishermen, the petitioners argued.

Commercial fishing interests and fish managers respond that only a small percentage of the kings intercepted in Bering Sea nets are on their way to Western Alaska rivers.

Returns of kings into the Kuskokwim and Yukon systems combined peaked at about 600,000 salmon a year, then declined in recent years to the new low average of half that, or 300,000 kings a year, according to a 2013 report by the Arctic-Yukon-Kuskokwim Sustainable Salmon Initiative, an effort that includes the tribal organizations, state and federal governments and fishing interests.

The accidental catch only would have been a major factor if it were in the magnitude of 100,000 kings, the report said. Instead, it averaged 15,000 a year from 2008 through 2012 in the pollock fishery, and not all of those were headed to Alaska, the report said.

"Thus, this part of the domestic fishery cannot account for the striking decline in Chinook salmon abundance or even for a substantial proportion of the decline. Clearly, other sources of mortality must also have contributed to the decline," the 2013 research action plan said.

The high seas bycatch is targeted because it's a factor that people can control, said Dawson Hoover, communications manager for Coastal Villages Region Fund, a seafood operator that runs a processing plant and a commercial operation on the Kuskokwim River and has interest in fishing boats including ownership of a Bering Sea factory trawler. That catcher-processer, the Northern Hawk, generated $50 million in sales in 2012, revenue that subsidizes the money-losing Kuskokwim commercial fisheries to the tune of $3 million to $5 million a year, Hoover said, citing Coastal Villages' 2012 annual report.

"The whole point is to provide jobs and commercial fishing opportunities where people can earn money, where they may not be able to make any money elsewhere," Hoover said.

Still, in 2007 the accidental catch of kings topped out at more than 121,000 fish. Tribal groups don't want to risk a spike in bycatch in a year where village residents went without, Naneng said.

"Bycatch is not the only problem, but they should be part of the solution," Naneng said.

The North Pacific Fishery Management Council next meets in October in Anchorage. Naneng says he hopes the petition seeking quick action is addressed. The council can't adopt emergency regulations but could recommend a course of action to the commerce secretary, said David Witherell, council deputy director.

5.0. Extant Knowledge as the Baseline[2]

The practices used to ensure the sustainability of fisheries in the Bering Sea are not unique. Even more, the preexisting regulation of the fishing industry points to a specific variant of a generic problem well known to social scientists. On the one hand are actors' particular interests, and on the other is the common good. Consider first the individual interests. Today commercial fishing provides millions of jobs and produces billions of dollars in revenue in the world annually; it figures prominently in international trade, as billions of dollars' worth of fish are traded across the globe each year (Food and Agriculture Organization 2014). Millions of small-scale fishers worldwide rely heavily on fishing for their livelihood; fish is a key source of food and protein for small-scale farmers as well (e.g., Allison et al. 2012; Teh and Sumaila 2013). Moreover, globally millions of people participate in recreational fishing. In the United States alone, recreational fishing annually generates about $30 billion (Southwick Associates 2006). Hence many people depend on the world's fisheries for subsistence and income. On the face of it, these individual interests should aggregate into a collective goal of protecting the sustainability of the world's fisheries.

Yet many of the world's fisheries are at risk due to human behavior. Economists and political scientists have long recognized that, while every fisher[3] involved might share a collective best interest to make sure that a fishery remains viable, the individual rational choices of fishers will bring to them to overfish. Without any property rights over any of the fish in a fishery until the fish are caught, fishers feel the need to fish as quickly and as much as possible: they fear that other fishers will try to claim the fish before them. Fishers thus engage in a race to fish, which leads to overfishing (e.g., Costello, Gaines, and Lynham 2008; Edwards 2003; Grafton et al. 2006; Hardin 1982; Olson 1965; Ostrom 1990). Overfishing in turn makes a fishery unsustainable, as not enough fish survive to produce a similar fishing population in subsequent seasons. As a consequence of this behavior across the globe, the biomass of the ocean's valuable and predatory fish (such as cod, tuna, or shark) has decreased by 90 percent in the last fifty years. Moreover, in some fisheries, catch per hour has decreased dramatically over the same span (Agriculture and Rural Development Department 2004).

Regulation addresses the Tragedy of the Commons. Specifically, regulators use quotas to put a cap on the amount of fish a company can fish from a specific species. If a company overfishes that species, it can be fined. This way, individual fishers need not fear that other fishers will overfish before they do, leaving them without fish in the foreseeable future. With the

property rights thus entailed, fishers know that no fishers other than themselves are allowed to catch their fish (as long as regulations are enforced).[4]
In addition, because fishers have a property right over future fishing stocks,
as long as they do not need fish for subsistence in the immediate future,
these fishers have a vested interest in maintaining the sustainability of the
fishery so that they can fish over the long term (e.g., Allison et al. 2012;
Beddington, Agnew, and Clark 2007; Costello et al. 2008; Edwards 2003;
Gibbs 2009; Grafton et al. 2006; Ostrom 1990).

As a precursor to drawing narratives from this story, we need to establish understanding of the status quo regulation and the rationale for it. The
story suggests that, without regulation, commercial fisheries will overfish.
Commercial fisheries have at their disposal advanced technologies that
enlarge their catches. Commercial fisheries aim for profit in a competitive
marketplace, and they claim the benefit of providing jobs. When regulation
is in place to prevent the depletion and even the destruction of the collective resource of fish, it acts as a binding constraint on commercial fisheries'
utility maximization, other things equal. For their part, as commercial fishing companies operate, they are as fully aware of the underlying problem as
are their governmental and intergovernmental regulatory overseers.

Analysts have long known that underlying problem as the Tragedy of
the Commons (e.g., Feeny et al. 1990; Hardin 1968, 1998). Given accumulated scholarship, we are justified in saying that implemented status quo
solutions to the Tragedy of the Commons, as well as anticipated behavior and outcomes without such solutions in place, constitute the structural
context within which every individual actor is embedded. Adopting terminology introduced by Tsebelis (1990), whatever happens in the commercial
fishing industry, or whatever regulators do, or whatever fallout communities experience, all of that is "nested" within the structural environment of
the industry-wide Tragedy of the Commons. The abstraction of the Tragedy of the Commons becomes the shared mental construct that creates the
context for actors' subsequent behaviors.

The Tragedy of the Commons is no longer a puzzle, then: it rather
enters into the baseline knowledge that all social scientists share. Even so,
to demarcate explicitly what does not and does require detailed examination, we begin with the standard model for the Tragedy of the Commons:
the Prisoners' Dilemma. The Prisoners' Dilemma can serve as a ready
module, as it forms a trusted part of the stock of existing knowledge. We
need not reinvent or explicitly model its logic as we proceed with narratives
and models. Whereas the news story describes overfishing and the necessity of regulation as the larger factual context in which the events unfold,

the Tragedy of the Commons and thus the Prisoners' Dilemma furnish the larger *theoretical* context for all our narratives and models, despite their individual theoretical biases. Insofar as we limit models here to the rational choice approach, the PD is the shared extant knowledge in which modeling choices must be anchored.

Indeed, given our premise that actors as well as analysts subscribe to the same theoretical construct, we can take the baseline module of the PD as a not-so-black box whose outcome is fully predictable and is common knowledge. Thus, we can assume that not only we as analysts but also every actor involved in our narratives anticipates the output generated in that box: namely, all will suffer and the resource will be destroyed unless effective regulation and effective enforcement are implemented. Our current point is that we have no reason to raise the question of whether actors would engage in costly cooperation if left to their own devices. That the model for the Tragedy of the Commons forms part of common knowledge means that we already know the answer: it is a resounding "No." We can also presume that we know why actors on their own do not cooperate: it is a dominant strategy to defect, as in the PD. In this way, we build on past formal models and need not complicate the design of our current models by rehashing past ones: we take the past models' *predictions* as known constants and incorporate them directly into the new models we fashion. We contribute to the accumulation of knowledge, confident that the ready-made preexisting components are logical and logically replicable, and that their underpinning assumptions are fully taken into account. We thus use accumulated knowledge and in doing so we produce models capable of further building on that knowledge. That is, because our new models connect to accumulated scientific knowledge by their assumptions and logicality, we add to the stock of accumulated knowledge as well.

Modeling the Prisoners' Dilemma becomes the basis for the original models we subsequently develop, as inspired by this story and the multiple narratives we pull from it. Recognizing that the abstraction is so firmly ensconced, we translate the PD into the world of fishers and fisheries, a world in which without regulation individual interests overwhelm shared interests, that is, the Tragedy of the Commons (e.g., Ostrom 1990).

Consider figure 5.0.1, depicting the Prisoners' Dilemma in which each of the two fishers, Fisher 1 and Fisher 2, gets to decide how much to fish without outside supervision. Each fisher has two choices: to overfish, L, or to limit her fishing, $L/2$ (i.e., to cooperate). If they both fish $L/2$, the fishery is sustainable. When only one abides by the quota ($L/2$, L) or (L, $L/2$), the

		Fisher 1	
		L/2	L
Fisher 2	L/2	10, 10	5, 15
	L	15, 5	7, 7

Key:
If {L/2; L/2}, then each catches 10
If {L; L/2} or {L/2; L}, then violator catches 20, complier catches 10, and both suffer the loss of 5 from the depletion of the fishery
If {L; L}, then each manages to catch only 17, and since the fishery is excessively depleted each also suffers the associated loss of 10

Figure 5.0.1. Extant knowledge: Members of the commons without regulation

violator catches twenty, but they both lose five in a public bad of depleting the fishery. When both overfish, (L, L), neither can now catch the full twenty. Instead, each catches seventeen; each also loses ten from the public bad of extreme damage done to their shared resource.

But each individual fisher would receive a higher payoff if both chose to cooperate with the other in protecting the shared resource, as in the top-left cell $(L/2, L/2)$. Given the fact that they would be better off if they both chose not to overfish, fishers in these situations have a collective interest to turn to fishery management techniques and impose coercive rules on themselves and also on everyone else who fishes their fisheries. They voluntarily create coercive solutions to the Tragedy of the Commons (Olson 1965).

Agents themselves support the creation of coercive mechanisms in the form of regulations. These coercive mechanisms improve the outcome, changing it from that of resource depletion to the one of cooperation and resource sustainability (Allison et al. 2012; Beddington, Agnew, and Clark 2007; Costello, Gaines, and Lynham 2008; Gibbs 2009; Grafton et al. 2006; Olson 1965; Ostrom 1990). It is important to note that in the story here fishers openly acknowledge that they support the overall practice of fishery management by regulators. Figure 5.0.2 illustrates how creating coercive regulations transforms the game, so that it is no longer a Prisoner's Dilemma and cooperative behavior becomes individually rational.

The game in figure 5.0.2 is similar to figure 5.0.1, with one exception: fishers are monitored. Now, if a fisher overfishes (plays L), she automatically pays a fine of F (gains twenty for the fish caught, loses five for the unsustainability of the fishery, and then also pays F, the fine; we assume

Fisher 1

	L/2	L
Fisher 2 L/2	10, 10	5, 15-F
L	15-F, 5	7-F, 7-F

Figure 5.0.2. Extant knowledge: Members of the commons opt for regulation

Fisher 1

	L/2	L
Fisher 2 L/2	**10, 10**	5, 9
L	9, **5**	1, 1

Figure 5.0.3. Extant knowledge: Members of the commons opt for regulation and establish the level of punishment F = 6

for simplicity perfect oversight with no cost). The rest is as before: if each fisher fishes to $L/2$, they each get a payoff of ten (for the fish caught). If one fishes to L and the other fishes to $L/2$, the fisher who fished to L gets fifteen minus F (twenty for the fish caught, minus five for the fishery's unsustainability, and minus F for the fine). The fisher who fished to $L/2$ gets five (ten for the fish caught minus five for the unsustainability). If both fishers race to fish, they both get seven minus F (seventeen for the fish, minus ten for depleting the fishery, and minus F for the fine).

Note that not just any level of punishment would ensure cooperation as individual choice. In order to ensure that no matter what the other fishers are doing, each fisher chooses to cooperate and only fish to a sustainable amount, the appropriate fine for overfishing should be greater than the maximum value of F for which the following hold: $10>15-F$ and $5>7-F$. The fine should therefore be greater than 5.[5] As figure 5.0.3 reveals, given this fine, the fishers now have the incentive to cooperate and fish a sustainable amount.

In the real world, fishers and regulators carefully design regulations such that they choose the optimal amount of fishing quotas, oversight, and fines for overfishing to ensure that fishers do not overfish. In the news story at hand, the existence and legitimacy of the regulator are neither puzzling nor contested among all those involved. The necessity and legitimacy of regulation are manifest shared knowledge for the real-world actors, just as for analysts. Neither do the fishers contest in any way the levels of quotas and fines when it comes to pollock fishing, which is the staple of the status-quo regulation. There is, however, a debate among fishers and regulators in the story regarding the best way to sustain the Chinook, which is harmed during pollock fishing. This concern for Chinook is superimposed on the resolved issue of preserving pollock fisheries in the Bering Sea.

5.1. Negative Externalities to Agents Outside
a Regulatory Regime

Our story highlights this additional problem in fishery management: the management of bycatch, or the accidental catch of nontargeted species while fishing. As the story suggests, commercial pollock fishers often catch Chinook salmon while fishing. The regulations that govern the management of pollock fisheries in the Bering Sea apparently do not guarantee that Chinook fisheries also remain sustainable, and thus threaten the livelihoods of subsistence fishers who rely on Chinook. This feature of the story prompts the first query we explore. That is, our *Query 1* motivating the first narrative is: How can regulation, as the actors support it and the story reports it, coexist with the costs of the bycatch suffered by the local communities?

5.1.1. Technology Choice to Minimize Costs Is Costly for Others

The Prisoners' Dilemma reveals why commercial fishers would overuse a fishery if left to their own individual, unfettered initiative. Extant regulation prevents them from doing so. Beyond that, why would they use fishing methods harmful to other species (or, more generally, harmful to the environment)? In particular, why would they decide to own and use a factory trawler, as named in the story? A factory trawler is equipped to process fish onboard and to stay at sea for weeks at a time. Technologies for bottom trawling (using large nets that scrape the floor of the sea) are especially crippling to ecosystems and species, yet midwater trawling, used in harvesting pollock, also brings grave damage (e.g., Hillers 2016; Scheer and Moss 2017).

In the instance of the story at hand, the fishing technology that commercial pollock fishers use generates the accidental bycatch of Chinook salmon of such magnitude that it threatens local Alaskan communities. In the context of the theory of externalities for outside agents, we construct Narrative 5.1.

Narrative 5.1: Actors in a Resolved Tragedy of the Commons Still Generate Negative Externalities for Outsiders

... Commercial fishing interests and fish managers respond [to the petitioners for the emergency cap on bycatch] that only a small percentage of the kings intercepted in Bering Sea nets are on their way to Western Alaska rivers.

Returns of kings into the Kuskokwim and Yukon systems combined peaked at about 600,000 salmon a year, then declined in recent years to the new low average of . . . 300,000 kings a year, according to a 2013 report by the Arctic-Yukon-Kuskokwim Sustainable Salmon Initiative. . . . The accidental catch only would have been a major factor if it were in the magnitude of 100,000 kings, the report said. Instead, it averaged 15,000 a year from 2008 through 2012 in the pollock fishery, and not all of those were headed to Alaska, the report said. ". . . This part of the domestic fishery cannot account for the striking decline in Chinook salmon abundance or even for a substantial proportion of the decline . . . ," the 2013 research action plan said.

The high seas bycatch is targeted because it's a factor that people can control, said Dawson Hoover, communications manager for Coastal Villages Region Fund, a seafood operator that runs a processing plant and a commercial operation on the Kuskokwim River and has interest in fishing boats including ownership of a Bering Sea factory trawler. That catcher-processer, the Northern Hawk, generated $50 million in sales in 2012, revenue that subsidizes the money-losing Kuskokwim commercial fisheries to the tune of $3 million to $5 million a year, Hoover said, citing Coastal Villages' 2012 annual report.

"The whole point is to provide jobs and commercial fishing opportunities where people can earn money, where they may not be able to make any money elsewhere," Hoover said.

Aggressive fishing technology such as a factory trawler thus creates potential negative externalities beyond the original Tragedy of the Commons that impinged on pollock. The narrative presents us with the moving parts for answering the theoretical puzzle: How can actors in a resolved Tragedy of the Commons still generate negative externalities? The answer implied by the narrative is that commercial profit maximization under regulatory fishing quotas drives the use of "efficient" and currently permissible aggressive technology. Such aggressive technology creates negative externalities by harming nontarget species on which others rely, affecting communities outside the original circle of stakeholders for the extant regulation. These are the communities that mobilize and announce their involvement, petitioning to amend—that is, tighten—the regulation so as to protect their interests as well.

In the broader structural context of the resolved Tragedy of the Commons, we know that regulation limits the quantity of the catch of any individual fishing enterprise. Under such a constraint, the profit-maximization goal of commercial fishing can only be pursued by means of minimizing the cost per unit of catch, that is, via the choice of fishing technology. While aggressive technology can make the monetary cost of staple fishing per unit decrease, the cost to the environment of using such equipment may increase. These methods could gravely harm ecosystems around the globe, further undermining the vitality of fisheries as a renewable resource (Agriculture and Rural Development Department 2004; Edwards 2003; Gibbs 2009; Grafton et al. 2006; Pikitch et al. 2005).

To place this scenario in the context of our story, commercial pollock fishers should prefer to use environmentally unfriendly equipment as long as that equipment is more cost-efficient. To make this scenario concrete: commercial fisheries operate factory trawlers using, we assume, midwater trawling, as is typical for Alaskan pollock (Agriculture and Rural Development Department 2004).

If we were to suppose that such practices damaged the habitat for the same species of fish already protected by the regulator, the regulator could include technological constraints and bans as part of the set of imposed restrictions, in accordance with the existing mandate. In accordance with the logic of the Tragedy of the Commons, fishing companies then would support such additional regulation in general, because they would be averse to lowering the probability of survival of a given fishery and even of multiple fisheries due to environmental damage, just as when they are threatened by overfishing.

Our current narrative, however, features two types of actors, and these actors do not live off the same "commons"! While the commercial fisheries are catching pollock fish, and catch Chinook salmon only by accident, the subsistence fishers are catching Chinook. That is, the first actor, commercial fisheries, generates a negative externality for the second actor, when its fishing technologies generate excessive Chinook bycatch as an unintended consequence.

Thus, the conflicts in the Alaskan story and in Narrative 5.1 link two tiers of fishing, commercial fishing and subsistence fishing, which are conducted in different places, by different actors, using different technologies, and focusing on different species. Tribal governance groups voice the concerns of subsistence fishers here. In petitioning for the emergency cap on Chinook bycatch, the governance groups focus on the byproduct of the commercial fisheries' use of advanced technologies for fishing pollock.

A complication for modeling is that the negative externality to subsistence fishers occurs probabilistically (i.e., breeding populations of Chinook may or may not return to spawning grounds in sufficient numbers), and this probability increases as the bycatch of the commercial fleet increases. Thus, even if we assume, as we do in Model 5.1, presented just below, that Chinook decline can occur only under conditions of high bycatch, when high bycatch *does* occur, the depletion of Chinook may or may not result. According to interested actors, that is, commercial fishers, this undermines the cause and effect argument for technologies that are less destructive to Chinook.

5.1.2. Model 5.1: Not in Common!

A model such as that in figure 5.1, when it accompanies a theoretical argument, serves primarily as an illustration, for it does not reveal any relationships in addition to those already explicated. That is, the theoretical claim is straightforward, and does not involve the multistep logic that would require formal verification. The very stage of putting the model together, however, forces the analyst to treat the commercial fishery as the key decision maker here, and to begin by specifying the preferences of that decision maker, rather than those of the affected subsistence fishers.

The model that we design from this narrative is one in which the decision maker, a commercial fishing Business, makes a choice of a technology, which leads to increased risks and thus lowered expected payoffs for a subsistence fisher. Observe in figure 5.1 that a fishing Business, B, chooses between Conservative, C, and Aggressive, A, technologies. After that, excessive bycatch of Chinook either occurs or does not occur, randomly. We express these contingencies as a move by "Nature," N, a nonplayer, to indicate its chance occurrence. In line with our theoretical discussion above, the probabilities with which Nature "chooses" high bycatch differ under different technologies. We denote those probabilities in figure 5.1 as $\alpha(C)$ and $\alpha(A)$, and specify that $\alpha(A) > \alpha(C)$, that is, high bycatch is more likely under the Aggressive technology.

Moving further down the decision tree in figure 5.1, we reach the stage at which the fate of the Chinook population is decided. For simplicity, we choose to assume that when bycatch is low, Chinook survival is assured (occurs with probability 1). Thus, the corresponding branches in the decision tree end with terminal nodes. Alternatively, as the story tells us, when bycatch is high, Chinook depletion may or may not result.[6] To capture that idea, we depict high bycatch branches as leading to further moves by "Nature," where Chinook survives with probability $\beta > 0$, or is depleted with probability $1 - \beta > 0$, as in figure 5.1. These new branches lead to terminal nodes as well.

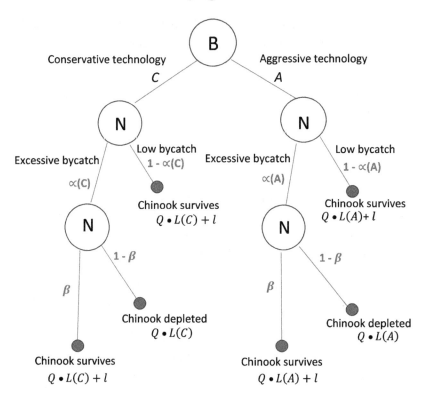

Figure 5.1. A commercial fisher's choice of technology

Terminal nodes in the decision tree in figure 5.1 indicate arrival at the outcomes, and our decision maker, the commercial fishing Business, has preferences over these outcomes. Thus, finally, we must assign the payoffs that the fishing Business stands to receive from each of the outcomes— only then will we know which decision, C or A, it will make. In order to assign payoffs, we need to specify the utility function of our decision maker.

In specifying a utility function, the modeler needs to identify the level of detail with which to characterize the utility that the decision maker in question derives from the feasible outcomes. In other words, the modeler has a choice of whether and how to think about the source of the decision maker's preferences. Technically, it is fine to simply assume that the decision maker prefers one outcome to others, and thus chooses the path that leads to that outcome. Yet doing so would leave us with little value added from the model and would approach a tautology, in which we assert that the actor chooses the action she prefers. In such a circumstance, the modeler would see her treatment of the decision maker's preferences as a lost opportunity.

First and above all, the fishing business wants to maximize its profit. As noted, the Tragedy of the Commons as resolved by extant regulation is *not* part of the model per se, but it informs the assignment of payoffs to the decision maker. Because we know that the regulator limits the quantity of pollock catch, we also know that the profit-maximizing business will fish at that limit, and will strive to increase its profit margin by the only method remaining to it: the choice of technology.

To amplify, since the business cannot increase output and thus reduce marginal costs, it must work to reduce the overall costs of the fishing operation, that is, to select cost-minimizing technology. The profitability of its chosen technology will be multiplied by the quantity as set by the quota limit in the business's utility function.

We elect to make an additional assumption that fishing businesses do not like the prospect of damage to Chinook populations via bycatch. We thus allow for a preference on the part of Business to preserve Chinook, but not an overwhelmingly strong one, as Chinook depletion does not harm its main resource base (pollock). Incorporating the assumption in the model emphasizes that the Business's driving preference for more profit overpowers its weaker preference for species preservation and pushes it to choose aggressive technology.[7]

Observe from figure 5.1 that excessive bycatch of Chinook salmon reduces resource availability for subsistence fishers only probabilistically. We calculate the probability of Chinook survival by multiplying the probabilities assigned to the branches for the moves by Nature, which lead to the outcomes in which Chinook survives, and adding up the probabilities of such outcomes. This probability we denote as $\chi(T)$. The commercial fishing Business's utility function then looks as follows:

$$u_b(T) = Q\,L(T) + l\,\chi(T), \qquad\qquad (5.1.1)$$

where Q stands for the quantity of pollock (note that Q is capped by regulatory quotas, and can be assumed to be constant, $Q = Const$, and equal to the quota for a profit-maximizing fishing business), χ stands for the probability of the local Chinook population's survival, $\chi(T) = \alpha(T) \times \beta + 1 - \alpha(T)$, T is fishing technology $(T = A, C)$, and L and l are coefficients.[8] $L(T)$ indicates the profitability of pollock fishing, with $L(A) > L(C)$, that is, fishing when using Aggressive technology is more profitable than fishing when using

Conservative technology. As stated, χ denotes the probability of Chinook survival: it is less than 1 if high bycatch occurs; if high bycatch does not occur, $\chi = 1$. The expression $l\,\chi(T)$ captures the probability of Chinook survival for a given choice of fishing technology. On this basis, we assign payoffs to the terminal nodes in Model 5.1, depicted in figure 5.1.

We solve now for the decision maker's preference over which choice to make, that is, which technology to adopt. Note that the probability of excessive bycatch α in figure 5.1 increases as the technology of commercial fishing changes from Conservative to Aggressive, that is, $\alpha(A) > \alpha(C)$ | $\alpha(T) \in [0, 1]$. Hence the implications of technology choice for the survival of Chinook can be understood as follows: $\chi(T) = \beta\alpha(T) + 1 - \alpha(T))$, or, equivalently,

$$\chi(T) = 1 - \alpha(T)(1 - \beta),$$

where β, again, is the probability of Chinook survival, conditional on high bycatch. Observe that χ decreases in α and increases in β.

The other part of the Business's utility function is not affected by the moves of "Nature" and remains either $QL(C)$ or $QL(A)$ for the technology choices of C or A, respectively, regardless of the fate of Chinook. Thus, we can now roll back in the decision tree and account for the probabilities of the moves of "Nature." The expected payoffs to the decision maker from each of its choices are as follows:

Technology choice is "Conservative" $\quad u_b(\alpha,\beta|T{=}C) = QL(C) + l\,\chi(C)$
Technology choice is "Aggressive" $\quad u_b(\alpha,\beta|T{=}A) = QL(A) + l\,\chi(A)$

The utility gain to a commercial fishery from choosing "Aggressive" technology is then

$$\Delta u_b = (\alpha,\beta|A) - u_b(\alpha,\beta|C) = Q(L(A) - L(C)) + l\,[\chi(A) - \chi(C)].$$

That is, it trades the increase in profits for a decrease in the probability of Chinook survival. Given our assumptions about the relative importance of the decline in the cost of pollock fishing and about the relatively low commercial fishers' loss from marginally increasing the probability of Chinook depletion, Δu_b above is positive, that is, $Q(L(A)-L(C)) > -[l(\chi(A) - \chi(C))]$. By assumption, the term in brackets is negative, therefore the right-hand side is positive; but it is of small magnitude. Hence the model illustrates that commercial fishers *will* make the decision in favor of Aggressive technologies.

5.1.3. Testable Implications from Model 5.1

We now draw implications from Model 5.1 that "convert" into observable behavioral patterns what we have just shown formally. First note that, even due to the specifics of the regulation, those actors who are regulated behave in ways that generate significant negative externalities beyond the regulated resource. This implication resonates with the abundant literature on regulation, which highlights the need for awareness that regulated industries can have broader external environmental impacts (e.g., Agriculture and Rural Development Department 2004; Beddington, Agnew, and Clark 2007; Branch 2009; Edwards 2003; Gibbs 2009; Grafton et al. 2006; Hedley 2001; Pikitch et al. 2004; Squires et al. 1998).

H5.1.1: Where quotas on producers are imposed, commercial fishers should select technologies that minimize the cost for that exact maximum quantity (e.g., fishing quotas).

This first hypothesis, recalling the model and our discussion of it, logically implies that over time, under the conditions of (H5.1.1), and if some technologies are banned, alternative technologies with cost-effectiveness optimized for the permitted quotas of catch would become available. This conjecture, however, is not immediately falsifiable. Nonetheless, it generates a corollary that is testable.

H5.1.2. Stable regulatory provisions should lead to technological change in the regulated industry.

Thus, the same narrative and the same model based on it generate several implications, some of which can, and others of which cannot, be subjected to empirical evaluation. In this section, we have used the resolved Tragedy of the Commons to pose the first query, built a narrative to address that query, and then designed a model. We now repeat the same process with a new query: How can subsistence fishers encourage commercial fishers to limit their bycatch?

5.2. Proposing a Change in Regulatory Provisions to Best Advance a Policy Goal

In the story at hand, the subsistence fishers affected by commercial fishers' technological practices propose a regulatory modification that does *not* involve the fishing quotas so familiar in the fishing industry and its regulation. From the subsistence fishers' request to lower the cap on bycatch, we discover that extant regulation already takes into account the impact of commercial fishers on Chinook, and already uses measures in addition to quotas to safeguard that species. Subsistence fishers focus on what is termed a "hard cap" on Chinook, not the quota on pollock, and advocate a more restrictive hard cap. As the second element of their petition, they call for a reduction in the bycatch threshold that triggers closer monitoring. Our next query, along with the narrative and the next model it inspires, explores the instrument of triggers to monitoring while emphasizing also the constraint created by the hard cap.

5.2.1. Regulation as an Object of Design

What is our **Query 2**? Exposing our process lightheartedly, how about asking: How can subsistence fishers manipulate commercial businesses to move as they desire? Or, to rephrase, which instrument would modify the incentives of the commercial fishing industry so that they would safeguard subsistence fishers' interests? As we construct Narrative 5.2 and design Model 5.2, we ignore the goals and preferences of commercial fishers, and we adopt the worldview of a subsistence fisher. We are poised to deepen the analysis of the regulatory mechanism, but we do so at the cost of sacrificing part of the "bigger picture." We accept the trade-off in order to pursue Query 2 where it takes us, to Narrative 5.2 and then to Model 5.2.

Narrative 5.2: Subsistence Fishers' Proposal for Regulating Commercial Industry

. . . The tribal groups want the government to lower the hard cap on the accidental catch of kings during the lucrative Bering Sea pollock fishery from 60,000 to 20,000 for the rest of 2014. There's a lower bycatch number that triggers tighter monitoring, and the groups also want to see it dropped, from 47,591 to 15,000.

. . . A 20,000-king cap would only have been exceeded once in the last five years, so putting it in place won't unfairly restrict the Bering Sea fishermen, the petitioners argued.

. . . The accidental catch only would have been a major factor if it were in the magnitude of 100,000 kings, the [Arctic-Yukon-Kuskokwim Sustainable Salmon Initiative] report said. Instead, it averaged 15,000 a year from 2008 through 2012 in the pollock fishery, and not all of those were headed to Alaska, the report said.

. . . Still, in 2007 the accidental catch of kings topped out at more than 121,000 fish. Tribal groups don't want to risk a spike in bycatch in a year where village residents went without.

This narrative focuses on the specifics of what subsistence fishers want to change in extant regulation. Of particular interest is their request to lower the threshold for more stringent monitoring. The trigger for closer monitoring consists of reaching some preset number of Chinook, reported as already caught by commercial fishers. The subsistence fishers ask for a reduction in this preset number, currently at 47,591; they propose the new number of 15,000. We design the model so that it discloses how this seemingly minor adjustment in enforcement of the overall policy might resolve the subsistence fishers' problem: the threatened disappearance of their Chinook fishing stock.

It is the subsistence fisher who acts as a designer of a revised regulatory mechanism in the current narrative. We theorize, then, that what the subsistence fisher proposes serves that fisher's goals the best. We assume for the purposes of this analysis and on the basis of the story that the subsistence fisher's utility is equivalent to Chinook survival. Such utility can be most simply defined as 1 if Chinook is sustainable, and 0 if not. Mindful that survival and destruction are matters of probabilities, and that we previously denoted the probability of Chinook survival as $\chi(T)$, this assumption gives us the subsistence fisher's utility function as: $u_s(T) = \chi(T)$. Considering that high bycatch occurs with greater probability under Aggressive technology, $\alpha(A) > \alpha(C)$, and that section 5.1.1 showed that $\chi(T) = \beta\alpha(T) + 1 - \alpha(T)$, it follows that $u_s(C) > u_s(A)$. In other words, the goal of subsistence fishers in proposing regulatory change is to motivate the fishing businesses to adopt Conservative technology, that is, to reverse the outcome as predicted in the model in section 5.1.1. Thus, in our current model, the fishing business will once again be making a technology choice. But the subsistence fishers' proposal will alter the environment for—the constraints on—such choice. Under the proposed new regulatory mechanism, the prediction should change, and Conservative technology should be adopted.

Alongside the caps that trigger tighter monitoring are other, higher caps—hard caps—on the absolute number of Chinook netted as bycatch. Although we as analysts have no specific information on implementation at either the vessel or industry level, hard caps mean that fishing would be stopped when the expected bycatch projected from monitors' observations exceeds some preset level that regulators identify as unacceptable. If such sanctions were applied at the level of the entire industry, then the obvious collective action problem would arise. The industry, however, might have methods for ensuring compliance of individual vessels. If sanctions were threatened at the level of the individual vessel, then we would expect vessels to use costlier practices for avoiding bycatch under the ongoing monitoring regime.

To reduce Chinook bycatch, as subsistence fishers want, commercial fishers would have to adopt one or more of several changes: employ different nets; slow down so as to give Chinook time to escape; and use special lighting. Such measures to curb bycatch would impose additional costs on commercial fisheries, which means that the fisheries would undertake them only if they were offset by a shift in some other, presumably regulatory, costs that the fisheries face given excessive bycatch.

The narrative suggests that whenever Chinook bycatch limits (in the form of a hard cap) are reached, fishing for pollock must stop—and must stop, indeed, whether commercial fishers have reached their quotas or not. The purpose of the closer monitoring, activated at the lower threshold, is to pinpoint the exact moment when the higher, hard cap is reached and thus further fishing must stop.

It seems, then, that only until and unless the close monitoring kicks in, commercial fishers have incentives to rush to fulfill their quotas ahead of each other. Once the more stringent monitoring comes into effect, commercial fishing activity will have to slow down to allow the monitors to record, compile, and tally up the ongoing bycatch from various vessels, and eventually all fishing must cease. Besides, it is possible that stricter monitoring would most affect those commercial ships that qualify as the "worst offenders"—that take in a disproportionately high amount of bycatch.

In this way, tighter monitoring, activated at a lower threshold, would shorten the period of unmonitored fishing when rapid harvesting of a large quantity of pollock is feasible, and further impede fishing with aggressive technologies during the period when monitors are at work. Closer monitoring would also raise the probability of not completing quotas due to the ban on fishing above the hard cap, which now cannot be accidentally overshot since closer monitoring provides better information. All these effects of stricter monitoring lessen the benefits reaped by commercial pollock fishers in adopting aggressive fishing technologies as modeled in section 5.1.2.

As Narrative 5.2 reports, the petitioners contend that a hard cap of 20,000 Chinook "would only have been exceeded once in the last five years." Combined with the information that the current hard cap is 60,000, and the current threshold for stringent monitoring is 47,591, this means that close monitoring simply has not entered into the commercial fishers' experience on a regular basis and has had little or no effect on their previous choice of fishing technology.

5.2.2. Reassessing Technology Choice under Variable Monitoring Thresholds

What is the impact of the petitioners' proposal on the payoffs to the fishing business? We start by taking the utility function for commercial pollock fishers that we used in section 5.1.2:

$$u_b(T) = QL(T) + l\,\chi(T),$$

Yet here the situation surrounding the choice of technology becomes more complex. Before, an Aggressive technology was unequivocally more profitable than a Conservative one; now, with a possible shift in monitoring rules, we introduce an important conditionality. Without monitoring changes, Aggressive technology yields more profit than does Conservative. With monitoring changes, however, the difference in profitability is, if not reversed, then at least diminished. We can denote the difference in profitability under old monitoring rules, $\neg m$, as

$$L(A|\neg m) - L(C) = |b|,$$

with notation otherwise as in section 5.1.2. We denote the difference in profitability under the proposal for new, revised monitoring regulation, m as

$$L(A|m) - L(C) = d < |b|.$$

As just noted, whereas d might be still positive, favoring Aggressive technology, it has a lower value under the new proposed regulation than under the extant regime.

Furthermore, whereas earlier commercial fishers were assured the full quota of catch, now there is a positive probability that fishing might be stopped if monitors declare that the hard cap on Chinook has been reached. For modeling purposes, then, we can presume that monitoring should reduce the probability of overshooting the hard cap.[9] Figure 5.2.2 thus has the fixed quantity, Q, that equals the full fishing quota only where the strict monitoring does not kick in. Where strict monitoring happens, we use a variable quantity instead, $K(\neg m, m)$, the expected quantity of catch, such that

$$K(\neg m) = Q;\ K(m) \le Q.$$

Note that figure 5.2.2 incorporates an additional assumption that the use of Conservative technology never triggers strict monitoring or generates excessive bycatch. In a separate treatment, not pursued here, an analyst might wish to consider that it is not an individual fishing business's but the *industry-wide* use of Aggressive rather than Conservative technologies that leads to bycatch reaching the level that triggers closer monitoring and thus might spur the monitors' move to halt fishing. In that analysis, a scholar might model the introduction of the institution of stringent monitoring as the industry's own effort to adopt uniformly Conservative technologies in order to avoid arriving at the hard cap. Because in this chapter we confine models to decision making by a single actor, we do not deal with that extension. Other scholars could do so, however, incorporating the model we present here as one of their building blocks and trusting it as a replicable element in their design.[10] In our model, we simply substitute the fixed quantity Q, the full fishing quota, with the expected quantity of pollock catch, K, and do not deal with the matter of how an individual fisher's choice of technology might impact that quantity.

Finally, here, unlike in Model 5.1, we can ignore any possible inclination of commercial fishers to preserve Chinook as a species, because ignoring it creates a bias against our expected prediction from the current model. Whereas in the first query, narrative, and model, we sought to explain the commercial fishers' choice in favor of Aggressive technology, here we seek to show that regulatory change (change in constraints) pushes the commercial fishers to adopt Conservative technology instead. Given that expectation, any preference on the part of commercial fishers for protecting Chinook would introduce a favorable bias, and we do not want to rely on that bias in our predictions: that would undermine our construct.

Thus, the new utility function for commercial fishers is as follows:

$$u_b(T, m, \neg m) = L(T, m, \neg m) \times K(m, \neg m) \qquad (5.2.1)$$

As in section 5.1.2, commercial pollock fishers are deciding whether to use Aggressive or Conservative fishing technology. Their payoffs are modified by the regulatory change to earlier close monitoring, which slows down the harvest of pollock even when the technology enables fast fishing, that is, it is Aggressive. Thus $L(A \mid m) < L(A \mid \neg m)$. This regulatory change, then, reduces the profitability of Aggressive technology. The monitoring aims to halt fishing when the hard cap defining maximum bycatch is reached, so under close monitoring, the permitted actual catch of pollock is likely to fall below the full quota, again, $K(m) < K(\neg m) = Q$.[11] We assume

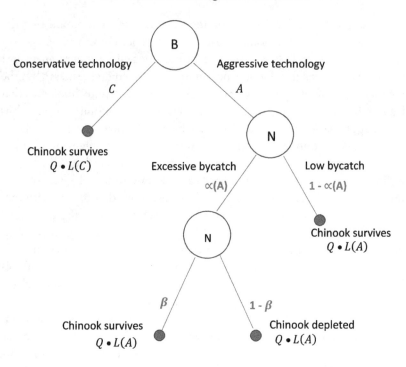

Figure 5.2.1. A commercial fisher's choice of technology under status quo regulation

that the choice of technology of an individual fishing business does not directly affect the probability with which the monitoring regime kicks in, μ. We also assume that under the status quo regulations, the probability of triggering the monitoring regime is zero, $\mu = 0$. That assumption is consistent with information included in Narrative 5.2.

With that in hand, Models 5.2.1 and 5.2.2 compare the decision environment for the commercial fisher before and after amending regulation to lower the threshold for stricter monitoring. Note that Model 5.2.1 is equivalent to Model 5.1 and represents the status quo regulatory environment.

5.2.3. Hypotheses: Instruments for Assessing Policy Experimentation

The testable hypotheses yielded by Models 5.2.1 and 5.2.2 are assessment instruments for policy effectiveness with regard to the presumed objectives of a regulatory regime. A key component of any fishing regulation is the

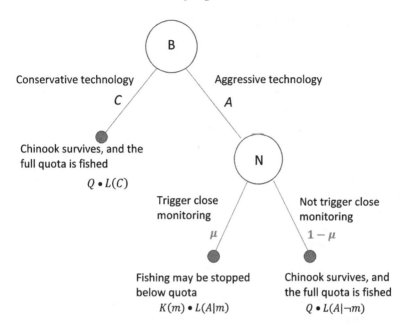

Figure 5.2.2. A commercial fisher's choice of technology under amended regulation

monitoring and punishment of fishers who cheat, for example, who poach, submit false reports, or disregard quotas. Without enforcement, we would return to the world before regulation, that is, the classic Tragedy of the Commons, with which we opened the chapter. Absent monitoring or punishment, fishers could still overfish without any immediate consequences (e.g., Allison et al. 2012; Beddington, Agnew, and Clark 2007; Costello et al. 2008; Edwards 2003; Gibbs 2009; Grafton et al. 2006; Ostrom 1990). With many vessels in operation belonging to multiple larger businesses, overfishing, or, in this case, exceeding the hard cap on bycatch, may even occur "accidentally," or unintentionally.

H5.2.1: The earlier the monitoring regime is in place, the lower the probability of exceeding a hard cap on bycatch.

We can operationalize the outcome so that it focuses on pollock rather than Chinook.

H5.2.2: The earlier the monitoring regime is in place, the lower the probability of fulfilling fishing quotas.

5.3. How to Regulate Use of Two Resources at Once?

Recall that Model 5.1, like the query and narrative inspiring it, focuses on the commercial fisher's choice of technology, a choice that has consequences for Chinook survival and thus for the subsistence fisher. Model 5.2 then incorporates the utility function of the commercial fisher, already developed in Model 5.1, as it shifts attention to the subsistence fisher and specifically to that fisher's expectation of gain from changing regulatory provisions. Now the regulator, absent throughout, reappears and moves to the forefront. We theorize that the regulator has to make its decision as if it were an agent of the two competing principals from the first two narratives.

The conflict between collective goals and individual interests requires regulatory effort. The conflict among the diverse individual interests of private actors jointly responsible for achieving collective goals requires even more regulatory effort. Diversity of interests can be rooted, for example, as we have seen, in different species of fish on which actors depend. Here, we also observe national, regional, and state actors beholden to different stakeholders and jointly engaged in the effort to regulate the catch of Chinook salmon, pollock, and other fish. Just as member states in international bodies may be concerned with different kinds of environmental impacts, federal agencies are likely to overlook the environmental impact that troubles regional agencies the most. The regulator here is regional, for the North Pacific Fishery Management Council (NPFMC) is composed of members nominated by North Pacific states and appointed by the secretary of the U.S. Department of Commerce, along with fisheries officials from North Pacific states (NPFMC 2017).[12]

All sides in this setting clearly concur on the decline in Chinook salmon. Yet they differ on what to do about it: whereas the petition advanced by the tribal governance groups calls for a new, lower cap on the bycatch, the commercial fisheries deflect the call, contending that the "bycatch is targeted because it's a feature that people can control." The claims and counterclaims point to a conundrum for a regional agency that must serve as an agent of these multiple, disagreeing principals.

5.3.1. How Hard Would a Subnational Agency Push for Regulatory Change?

Considering a contest between multiple principals over a joint agent charged with regulating their activity directs us to **Query 3**: Should the joint agent push for change that some principals want and others do not? If the agent were to push, then how strong should its push be? Specifically, in this story, should the Fishery Management Council advocate for regulatory change that may benefit subsistence fishers, but would inflict costs at least in the short run on commercial fishers (as in the comparison between Models 5.2.1 and 5.2.2)?

Narrative 5.3: The Fishery Management Council: Servant of Two Masters

The Association of Village Council Presidents and the Tanana Chiefs Conference filed their petition with the U.S. Department of Commerce secretary and the North Pacific Fishery Management Council for an emergency cap they say is needed to avoid substantial harm to the kings . . . and to communities up and down the Kuskokwim and Yukon rivers.

. . . The accidental catch only would have been a major factor if it were in the magnitude of 100,000 kings, the [Arctic-Yukon-Kuskokwim Sustainable Salmon Initiative] report said. Instead, it averaged 15,000 a year from 2008 through 2012 in the pollock fishery, and not all of those were headed to Alaska, the report said.

. . . "The whole point is to provide jobs and commercial fishing opportunities where people can earn money, where they may not be able to make any money elsewhere," Hoover [of the Coastal Villages Region Fund, a seafood operator] said.

Still, in 2007 the accidental catch of kings topped out at more than 121,000 fish. Tribal groups don't want to risk a spike in bycatch in a year where village residents went without, Naneng [of the Association of Village Council Presidents] said.

. . . The North Pacific Fishery Management Council . . . can't adopt emergency regulations but could recommend a course of action to the commerce secretary.

Given technological advances, the decision-making challenges facing regulatory bodies may be exacerbated by their own institutional design (Farrer, Holahan, and Shvetsova 2017). To appreciate this point, recall the composition of the NPFMC, and note too that a scientific committee advises it, as does a panel composed of different sectors of the fishing industry, recreational fishers, consumers, and environmentalists (NPFMC 2017). The NPFMC as a regulator thus comprises state-level actors, including some who must earn federal appointment following state nomination, and receives advice from the very stakeholders who stand to gain or lose from its recommendations. The agency is not exactly captured, because of the checks and balances between and among its conflicted stakeholders, but neither is it free. This hidden regulatory capture by multiple principals with conflicting interests is the essence of the model: we home in on the agent's balancing act as we design Model 5.3.

Can regulatory bodies such as this adequately balance the multiple policy goals of their diverse stakeholders? The existence of multiple policy goals and the need to address them all at once is acknowledged in extant literature. While fishery management techniques based on allocating specific quantities of catch to fishers, such as individual transferable quotas (ITQs), have had some success in ensuring the sustainability of staple fishing stocks (e.g., Costello et al. 2008), those techniques typically have managed only a single species. Policymakers, however, have long been aware of the problems arising from the use of techniques optimized for catching a target species when those lead to overfishing other species. In order to maintain the sustainability of multiple fisheries, scholars in multiple disciplines have called for taking into account the entire ecosystem from the start, even when trying to manage a single species (e.g., Beddington, Agnew, and Clark 2007; Edwards 2003; Grafton et al. 2006; Hedley 2001; Ostrom 1990; Pikitch et al. 2004).

The task confronting the regulators is a thorny one: to devise regulations that might prevent the depletion of both scarce resources (two species of fish), while protecting the individuals and collectivities whose livelihood depends on either one of those resources. Among the multifold concerns at play are the sustainability of Chinook salmon, yet also the profitable, indeed "lucrative," fishing business of commercial fisheries, the livelihoods of fishers, their families, and their communities, and the broader status of ecosystems and the environment.

Observe that the North Pacific Fishery Management Council, as a regional council, is charged with regulating only a regional subset of the national commercial fishing industry, and that local Native American subsistence fishers form another significant constituency. Recall that the Fishery Management Council does not create regulations on its own but it can "recommend a course of action" to the U.S. secretary of commerce. Consider that the office of the U.S. secretary of commerce, for its part, has a nationwide mandate and is unlikely to be immediately sensitive to the concerns of subsistence fishers in any given locality. The narrative suggests that the NPFMC cannot wait for the national regulator to act on this issue if it wants to shift the status quo, because that would amount to waiting too long, given its stakeholders' preferences. Thus, we design our model to evaluate how likely the Council is to recommend action to the secretary of commerce.

5.3.2. Balancing Stakeholders' Interests

The narrative does not disclose the eventual outcome: we do not know whether the rules were in fact amended. The news story implies rule amendment as a possibility. Recall our selection criteria for news stories, and recognize that this story meets the criteria, in that the entire sequence of actions remains to be played out and the outcome of it is uncertain. We design Model 5.3, then, so that it specifies the conditions under which the Fishery Management Council would recommend an emergency cap, as the petitioners propose, or uphold the status quo.

The model in figure 5.3 starts with the Fishery Management Council's decision on whether to issue a recommendation to change extant regulation or leave the status quo in place. We have both decision branches ending in terminal nodes in Model 5.3. The complexity of this analysis lies in assigning payoffs to the two outcomes and in comparing them to see under what conditions the decision maker would prefer one choice over the other.

In the modeled universe, the principals of the Council are the two stakeholders. In the real world, many more stakeholders may actually be engaged in its activities. The stakeholders in our model number only two because the conflict at issue concerns them, and their joint agent is balancing their preferences according to our theory. The two stakeholders relevant for this decision are the commercial fishers of pollock and subsistence fishers of Chinook. We can reasonably assume that some balance of the benefits to the two stakeholders represents the payoff to the Fishery Management Council. Without loss of generality, we can normalize the weights of the stakeholders' payoffs in the Council's utility function as γ, $1-\gamma \in [0, 1]$, where γ is the weight on the benefit to commercial fishers and $1-\gamma$ is the weight on the benefit to subsistence fishers. This assignment of utility weights would apply when the preferences of stakeholders do and do not align. In fact, as seen in section 5.1.2 above, commercial fishers choose to act so as to threaten Chinook survival even though we assume that, other things equal, they have a preference to preserve Chinook. In part, then, the commercial fishers' utility accords with that of subsistence fishers; yet the part in their utility that reflects a profit maximization goal works against the subsistence fishers' preference.

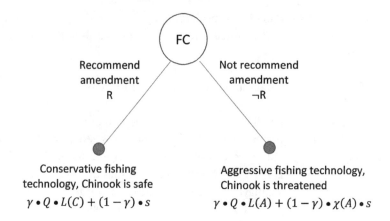

Figure 5.3. Balancing conflicting preferences when making a decision

The utility function that we assign to commercial fishers in section 5.1.2 serves our needs here as well, and is expressed as

$$u_b(T) = QL(T) + l\,\chi(T),\qquad(5.3.1)$$

where, as before, χ stands for the probability of the Chinook population's survival, which depends in turn on the probability $\alpha(T)$ with which excessive bycatch occurs, T is fishing technology ($T = A, C$), $L(T)$ is the profitability of commercial fisheries, with $L(A) > L(C)$, that is, Conservative technology is less profitable than Aggressive, and $\alpha(A) > \alpha(C)$, that is, Conservative technology is safer for Chinook since it reduces the frequency of high bycatch.

We face a presentation choice at this point: Should we incorporate the change that amendments to status quo regulation would cause in the profitability of the aggressive technology and in the expected quantity of catch? Doing so would increase the complexity of the model's design without changing its findings. We explain why in what follows.

First, we can safely use the constant Q, indicating the full quota of pollock, instead of the variable $K(T)$, which indicates expected pollock catch and accounts for a potential early stoppage due to exceeding the hard cap on Chinook bycatch. In fact, in light of the now solved Models 5.1 and 5.2.2, we have good reason to expect that a simplification from $K(T)$ to Q in the fishing Business's utility function would not affect the substantive dynamics in Model 5.3. This is because, viewing Models 5.1 and 5.2.2

as two possible contingencies dependent on the choice of regulation, as in Model 5.3, we include as payoffs the predicted play in each of them, respectively: the prediction from 5.1, where regulation is not amended, and the prediction from 5.2.2, where regulation is amended. In the former case, the full quotas are fished because there is no close monitoring; in the second case, the predicted choice of technology is Conservative, and so the hard cap is not reached and, again, the full quotas of pollock are harvested. In other words, and adopting, in preparation for chapter 7, the language of noncooperative game theory, *on the equilibrium path*, the catch is always Q. And, again using the terminology of noncooperative game theory, our Models 5.1 and 5.2.2 are proper subgames of Model 5.3—if instead of pay-offs at 5.3's terminal nodes we included the means by which those payoffs were derived in previous sections. By incorporating the predictions from 5.1 and 5.2.2 as payoffs in 5.3, then, it is as if we *rolled back* through Models 5.1 and 5.2.2.[13] The bottom line: the regulator in Model 5.3 need not worry about the quantity of catch parameter, as it does not vary as a result of her decision, because rationally responding commercial fishers will end up with their full quotas regardless.

Second, we do not need to indicate in the payoffs of Model 5.3 that the profitability of the Aggressive technology will be lowered in the case of amendment to the regulation as proposed. The reasons are the same as above: taking our prior Models 5.1 and 5.2.1 as building blocks and (new) extant knowledge, and using them in the model we now design, observe that (1) absent regulatory change, the technology choice *will* be Aggressive and the utility function in expression 5.3.1 is accurate, since no close monitoring or hard cap constraints will be encountered, and (2) with regulatory change, the technology choice *will* be Conservative, and we do not therefore need to worry about the decline in Aggressive technology's profitability. Expression 5.3.1 then will serve well to capture the utility function of the commercial fisher-stakeholder.

The utility function of subsistence fishers, u_s, we defined in section 5.2.1 as $u_s(T) = \chi(T)$. This is how the Council will perceive the preferences of subsistence fishers as its stakeholders.

Since the Council values the success of both commercial and subsistence fishers, we can now express its utility function as in Equation 5.3.2, presented just below. The Council is thus viewed as a servant of multiple masters, in accord with the Council's institutional design and as discussed. Nonetheless, in deciding on its recommendation with regard to regulating bycatch, the Council has to choose a side. Which way it decides would depend on the exact value of the parameters involved. The resulting utility of the decision maker in this model, then, is

$$u_{FC} = \gamma u_b + (1 - \gamma)u_s, \qquad (5.3.2)$$

More specifically, we now focus on the Council's own decision to recommend or not recommend the amendment to existing regulation, and express the Council's utility as

$$u_{FC}(T(m, \neg m), R) = \gamma(QL(T) + l\,\chi T) + (1 - \gamma)\,\chi\,(T),$$

where R indicates the Council's decision to recommend or not, and all other variables are as previously defined. If the Council issues a recommendation, that would increase the probability of new regulation, which would effectively limit commercial fishers to the use of Conservative technology as shown. Since by assumption the Council's recommendation increases the probability of regulation, the Council will issue a recommendation only if its utility does not diminish with a shift from Aggressive to Conservative technology on the part of commercial fishers, that is, if

$$\gamma Q[L(C) - L(A)] + \gamma l\,[\chi(C) - \chi(A)] + (1 - \gamma)[\chi(C) - \chi(A)] \geq 0 \quad (5.3.3)$$

Noting once again $\chi(C) - \chi(A) > 0$, while $L(C) - L(A) < 0$, expression (5.3.3) holds when

$$\gamma \frac{Q[L(C) - L(A)]}{\chi(C) - \chi(A)} + \gamma(l - 1) + 1 \geq 0, \text{ or}$$

$$\frac{Q[L(C) - L(A)]}{\chi(C) - \chi(A)} + (l - 1) + 1/\gamma \geq 0. \qquad (5.3.4)$$

Replacing for convenience the increase in the likelihood of Chinook survival when Conservative technology replaces Aggressive technology as $a = \chi(C) - \chi(A)$, and the decline in profitability from switching to Conservative from Aggressive technology as $b = L(C) - L(A)$, and using now a and b as substitute notations to simplify presentation, we transform (5.3.4) as

$$\frac{1}{\gamma} \geq 1 - Q\frac{b}{a} - l, \text{ or}$$

$$\gamma \leq \frac{1}{1 - Q\frac{b}{a} - 1} \qquad (5.3.5)$$

In expression (5.3.5) we now have a restriction on the parameter γ for the Fishing Council to want to change regulations and thus to issue its recommendation. It will issue a recommendation only if the weight of commercial fishers' benefits in the Council's utility satisfies (5.3.5)—that is, it is smaller than the right-hand-side expression.

Before we can conclude, however, that this is what is required for the Council to want to recommend the regulatory change, it is crucial to check the feasibility of the condition that we have derived and that needs to hold in order to observe target behavior. In other words, we need to check if the Council *ever would* prefer to make such a recommendation. In the terminology of chapter 2, we need to ensure that the *Existence* test is met.

Hence, we proceed to observe: mathematically, the weight of the commercial fishers' payoffs in (5.3.5), γ, needs to be positive, by assumption. That means that the right-hand side of the expression (5.3.5) must be positive, and so the denominator in it must be positive: $1 - Qb/a - l > 0$. Thus, before we can seek to identify when the condition for the agent's support of a change to the status quo can be satisfied, we need to specify the parameter domain where it is even possible for it to be satisfied. Given that it contains $-l$, such a domain could be empty, which would mean that the condition we derived would be a mathematical tautology. If that were the case, then we would land in Quadrant 3 in figure 2.2: we would be unable to ever explain the observed behavior in a model based on our current set of assumptions, for we would be violating *Existence*.

Checking 5.3.5 this way, we need to establish that the domain, the parameter range where it is satisfied, is nonempty, that is, that its right-hand side, *can* be positive. Note that b is the negative term as we defined it, while a is positive. That implies that $1 - Qb/a > 0$, and so $1 - Qb/a > l$ can be easily satisfied.

Now that we have established *Existence*, namely that it is, indeed, possible under some conditions for the Council to prefer to recommend a change to extant regulation, we focus on meeting the *Logicality* test. *Logicality* would require that we establish the parameter bounds for when exactly the Council would prefer change over the status quo. It follows from (5.3.5) that the greater γ is, the smaller the right-hand side expression can be while satisfying (5.3.5). From that, it further follows that the cumulative probability of Chinook demise, $\chi(a)$, would need to be greater, the greater γ is, for the expression to hold. To drive home the substantive meaning: when commercial fishers as a set of stakeholders are relatively important to the Council, the probability of Chinook demise needs to be relatively great in order for the Council to issue a recommendation for more regulation in the form of a lower cap on bycatch.

To summarize, we have derived the conditions under which the Council would recommend further regulation to the secretary of commerce. These conditions are as follows:

$$\begin{cases} \gamma \le \dfrac{1}{1-Q_a^{\frac{b}{a}}-l} \\ 1-Q_a^{\frac{b}{a}} > l. \end{cases} \qquad (5.3.6)$$

We have now established that we have met both the *Logicality* and the *Existence* tests: we are in Quadrant 1 of figure 2.2, and are ready to ask, how do we translate Model 5.3's findings into hypotheses?

5.3.3. Testable Implications

A reasonable question arising in the context of expression 5.3.5 is the following: What change in parameter values might bring about a change in the Council's choice from upholding the status quo to recommending a new regulation to the secretary of commerce?

The first and most obvious answer is institutional: the Council would be more inclined to follow the subsistence fishers' preference if the weight of that group in the Council's utility were to somehow increase. This prediction is consistent with the accountability literature, as well as with the emerging literature on the technological implications of political representation (e.g., Balalaeva 2012, 2015; Rosenberg and Shvetsova 2016).

H5.3.1: The greater the share of subsistence fishers' representatives on the membership of the Council's advisory panel, the greater the likelihood that the Council will recommend regulatory change, other things equal.

Yet attend to the ceteris paribus clause: if we hold the institutional effects constant, then something else must change in the Council's environment, broadly conceived. Presumably, because the Council has had a hand in maintaining the status quo, the Council has preferred the status quo up to the filing of the petition. Nonetheless, the Council's preferences may reverse, and it may be moved to act, if the assessed probability of Chinook demise increases.

H5.3.2: The probability that the Council will recommend regulatory change will be greater in the period immediately after the appearance of new information about an increase in the threat to Chinook survival than otherwise, other things equal.

Suppose that the petition submitted by the native governance groups reflects their insight into what is feasible as an agenda for policy change. That is, it is in the interest of the governance groups to gather information and propose caps that are at least close to being acceptable to the Council, in light of its many stakeholders' interests. The groups' petition would inspire action by the Council only if it gathers and conveys enough new information, in particular, about the threat level to Chinook survival, χ.

5.4. Conclusions

This chapter is grounded in a generic problem in political and social science—and in political and social life: the conflict between individual interests and collective goals. We have pursued distinct queries in each of this chapter's sections, as in chapter 4. Yet here, unlike in chapter 4, we have used the results of one model as building blocks in constructing subsequent models. Chapter 5 thus holds an illustration of how to use extant models as we *build theory*, and in so doing contribute to the *accumulation of knowledge*. The condition for being able to incorporate findings from an extant model into a subsequent one is adopting that extant model's assumptions. This we have done by reusing previously defined payoffs and utility functions: we have engaged in a relatively simple form of *theoretical replication*. Observe as well that in this chapter we have shown an example of what fulfilling the conditions of *Logicality* and *Existence* looks like in the context of a simple model.

Another helpful way of looking at what we have done in this chapter, very different from how we approached modeling design in chapter 4, is that one can bring several simple models into one larger model, as long as the assumptions stay consistent throughout. In fact, section 5.3 comes very close to doing that—without, however, directly combining the models. If we had done that explicitly, of course, we would have had multiple decision makers making choices in the same model. As we have seen, one set of decision makers sometimes seeks to influence others' behavior (as subsistence fishers do in section 5.2), or have payoffs linked to the behavior of others (as do both subsistence fishers and the Fishery Management Council in sections 5.2 and 5.3). Therefore, in a combined model we would need to treat actors not as mere decision makers, but as strategic players: the combined model would be a game. In chapter 7, we turn to examine games, where the attempt to anticipate each other's actions and influence others' behavior is modeled explicitly.

In this chapter, we have chosen to build a sequence of decision-theoretic models with a simple benefits-maximizing approach to predicting actors' behavior. A reader who is familiar with game theory would readily see that Models 5.1 and 5.2 are in fact proper subgames to Model 5.3, and that this chapter's cumulative result can be interpreted (understood) as developing subgame perfect equilibria via backward induction in a game with complete information. We address more complex noncooperative games with incomplete information in chapter 7. Here, again, we draw the reader's attention to the fact that many social interactions can be assessed by rolling back through distinct decision-theoretic steps, as long as we can reasonably assume that everyone knows everything about each other, and observes each other's moves: that is, everyone has complete and perfect information. Once more, to do that, an analyst must keep all assumptions consistent through all steps!

SIX

Setting the Agenda to Manipulate the Outcome

With Benjamin Farrer, *Knox College*

For the first and only time in our book, in this chapter we illustrate what would happen if the same narrative were modeled via alternative methods. Here, one narrative based on one story leads to the design of two distinct models—one a social choice model and the other a cooperative game-theoretic model. The object is to show that which method to use is truly a choice, an assumption. This choice, however, affects what predictions look like. In demonstrating this point, we also showcase the next two types of formal models from chapter 3.

The *Huffington Post* published the news story that supplies the empirical datum to launch this chapter's models. The event on which it reports was the September 2014 United Nations Summit on Global Warming—more specifically, a fairly brief speech delivered at the Summit by Prime Minister David Cameron of the United Kingdom (UK). The speech was unexpectedly controversial, so that it received separate coverage in media around the world, including in the *Huffington Post*.

Story:

Cameron Tells UN World Needs 'New, Ambitious Global Deal' to Restrict Climate Change

Anonymous

Huffington Post UK | 2014-09-23

Prime Minister David Cameron has called for the international community to come together behind a "new, ambitious global deal" to restrict global warming to 2C. But the PM insisted that there was no need to choose between economic growth and reducing carbon emissions, telling the UN Climate Summit that innovative technologies—including shale gas fracking and nuclear energy—could help rein in climate change.

Mr Cameron called for nations around the world to cut "green tape" bureaucracy which he said was providing perverse incentives to greenhouse gas-producing fossil fuels, while holding back clean new sources of energy. The Prime Minister was among 120 world leaders gathered at the Climate Summit called by UN Secretary General Ban Ki-Moon to pave the way for a global agreement on climate change at a conference in Paris next year.

Mr Cameron restated his backing for a 40% cut in greenhouse emissions by 2030 which is expected to be put on the table by the European Union. But he made clear that any deal in Paris must involve legally binding commitments for all countries around the world.

"We now need the whole world to step up to deliver a new, ambitious, global deal which keeps the two-degree goal within reach," Mr Cameron told the summit. "And I will be pushing European Union leaders to come to Paris with an offer to cut emissions by at least 40% by 2030."

Mr Cameron called for leaders to work hard to "raise the level of ambition" at Paris and prevent a repeat of the failed Copenhagen summit in 2009, warning: "We cannot put this off any longer. To achieve the deal we need all countries to make commitments to reduce emissions," he said. "Our agreement has to be legally binding, with proper rules and targets to hold each other to account. And we must provide support to those who need it, particularly the poorest and most vulnerable."

Mr Cameron said it was "completely unrealistic" to expect poorer countries to forgo the carbon-fuelled economic growth enjoyed by the West. But he insisted: "If we get this right there's no need for a trade-off between economic growth and reducing carbon emissions. . . . We need to give business the certainty it needs to invest in low carbon.

"That means fighting against the economically and environmentally perverse fossil-fuel subsidies which distort free markets and rip off taxpayers. It means championing green trade, slashing tariffs on things like solar panels. It means giving business the flexibility to pick the right technologies for their needs.

"In short we need a framework built on green growth not green tape." He added: "As political leaders, we have a duty to think long term. When offered clear scientific advice, we should listen to it. When faced with risks, we should insure against them. And when presented with an opportunity to safeguard the long term future of our planet and our people, we should seize it. I would implore everybody to seize this opportunity over the coming year.

"Countries like the United Kingdom have taken steps, we've legislated, we've acted, we've invested, and I urge other countries to take the steps they need to as well, so that we can reach this historic deal."

The summit follows massive demonstrations around the world last weekend, which saw celebrities including Leonardo di Caprio and Emma Thompson join thousands of activists in demanding action from world leaders on climate change.

Friends of the Earth's Campaigns and Policy Director Craig Bennett—who is attending the summit—said: "Arriving at a climate change summit with a speech that promotes fracking is like trying to sell cigarettes at a hospital.

"Twenty-first century problems need twenty-first century solutions: If we want to build a cleaner, safer future we must switch to renewable power and end our dirty addiction to fossil fuels. With clean renewable power becoming ever cheaper, available now and accessible to ordinary people, we simply don't need to frack. It's at best a red herring and at worst a dangerous folly.

"Warm words abroad come cheap, but success in the fight against climate change will be measured in concrete actions by leaders at home—and the Prime Minister's record leaves a lot to be desired."

Greenpeace UK climate campaigner Sara Ayech said: "David Cameron was right to go to the summit and back a binding global deal, but he has to follow this through with concrete action at home.

"Right now, Britain is burning growing amounts of coal just because it's more lucrative for the Big Six than using gas. This is damaging our climate, our health, and our energy security as we depend on Vladimir Putin's oligarchs for most of our coal imports.

"We need action not just words. Will Cameron now cancel billions in planned new subsidies for UK coal? And will he make good on a rumoured commitment to phase out coal emissions completely?"

© Huffington Post UK

This story, although it reports on a single action of a single person, discusses the welfare of multiple actors, touches on interactions among those actors, and relays points of view expressed by varied observers of the prime minister's speech. It amply meets our criteria for selection of stories, including the criterion of being mundane. Neither the timing nor the implications were special: the speech was not delivered at any sort of critical juncture in the history of climate policy, nor did it become a groundbreaking, pivotal point. The UN speech was a relatively everyday event for the prime minister and apparently for the public. Consider: by June 2018, a recording of it uploaded to YouTube had reached only 4,535 views.[1] In contrast, by June 2018, Cameron's October 2015 Conservative Party Conference address had reached over 70,000 views, and his final July 2016 Prime Minister's Questions had reached over 800,000 views.[2] The ordinariness of the speech in principle would allow us to dissect it for a number of theoretical purposes, although in this chapter, unlike in others, we do not travel this route for the sake of brevity. Rather than pursuing multiple theories, here we pursue a single theory but more than one methodological approach.

6.1. All in a Day's Work: International Diplomacy on the Prime Minister's Multifold Agenda

The story not only relates the address given by Cameron at the UN Climate Summit, but also highlights two particular aspects of the speech: first, Cameron's commitment to reaching a legally binding international deal on climate change as soon as possible, and second, his insistence that such a deal must facilitate rather than forestall economic growth. In this context, no single narrative from the speech is ex ante more important and interesting to model. While climate change is one of the major political issues facing the planet (e.g., Keohane 2015; Klein 2014; Tobin 2017; Urban 2015), and this speech directly pertains to that issue, the speech does not qualify as a decisive pronouncement on the subject. It contains no major climate change policy announcements and no radical departures from prior rhetoric. It does, however, encapsulate many different ideas in a single short speech. It addresses issues tied not just to climate change but also to partisan pressures, domestic lobbying, energy politics, international diplomacy, and still more. If we turn this story one way, then one research question is visible, but if we were to turn it another way, then another research question would emerge. Across the spectrum of visible research questions, the general question of "what to do about climate change" might appear to be the one that looms largest, but the alternative narratives are no less important or interesting from the perspective of the social scientist.

In this chapter, we focus on the element of international diplomacy. We begin by briefly describing how this narrative differs from the conventional "collective action problem" narrative often cited in analyses of international climate change diplomacy. Then we discuss how international diplomacy is affected by Cameron's use of issue linkage strategies in the speech. He connects his broader energy policy agenda to his goals on climate change, and so shifts the agenda for the international negotiations. He aims to ensure that these negotiations inject his energy policy goals into the conversation about climate change. Indeed, one interpretation of our single narrative in this chapter is that Cameron forces international participants to focus on the broader energy policy issues in order to direct their actions on the distinct issue of climate change. Cameron engages in what Riker (1986) labels heresthetic manipulation. Interestingly, he does so by broadening the scope of the narrative and inviting the actors-cum-analysts to work out the solution for themselves. He anticipates that what they discover as a solution will fit well with Cameron's agenda.

We use the term "climate change" as social and natural scientists typically do: to refer to the anthropogenic emission of carbon dioxide and other greenhouse gases, particularly since the Industrial Revolution, which has changed the balance of gases in the atmosphere. This phenomenon leads to more of the sun's energy being reflected back to the Earth's surface. It does not simply lead to uniform warming across the surface of the planet, but rather to a bundle of different effects. Some of these effects have, in all probability, already begun, and others are predicted to develop in the next few decades depending on future emissions patterns. These effects include heat waves, rising sea levels, changing rainfall patterns, agricultural disruption, changes in disease vectors, refugee crises, increases in the frequency and severity of extreme weather events, and many other undesirable outcomes (for an overview, see, e.g., Masson-Delmotte et al. 2018).

Climate change has acquired the status of a salient political issue and of course served as the main topic of Cameron's speech at the UN Summit. Since climate change is caused by the emission of greenhouse gases, international negotiations have focused on the issue of how to equitably and efficiently reduce those emissions. Most scientists agree that emissions have already led to a global average temperature increase of around 1°C, and that avoiding a 2°C rise should be a top priority. Even a 2°C rise would represent a huge global change, but any temperature changes above that point would likely result in outcomes that are severely catastrophic rather than somewhat catastrophic. Yet to restrict temperature changes to less than 2°C would require steep cuts in emissions, perhaps steeper than 90 percent (e.g., Hawken 2017; Monbiot 2006). Since reducing emissions is costly, most formal models of these negotiations assume that political leaders have attempted, understandably, to argue against cutting emissions in their own country, and instead to impose the costs of emissions cuts on the citizens of other countries.

Most members of the international community have indeed resisted setting steep reductions targets, with the European Union (EU) setting a target of 40 percent cuts by 2030, relative to 1990 levels, and the United States at the time setting a target of approximately 30 percent cuts by 2030, relative to 2005 levels. This has led to conflict with "Non-Annex I" (NAI) countries, that is, the disparate set of low- and middle-income countries that have historically contributed far less to global greenhouse gas emissions and that did not take on binding targets under the 1998 Kyoto Protocol, the most prominent effort before 2014 to reduce greenhouse gas emissions.

Formalizations of international climate change negotiations are dominated by models that deal with this conflict (e.g., Hovi et al. 2014; Madani 2013). There are of course many other aspects to the negotiations, including issues of varying capacity for emissions, varying responsibility for historical emissions, and varying conceptions of climate justice (e.g., Vanderheiden 2009), but most formal models focus on the "collective action problem" narrative (Olson 1965). All national leaders want emissions to be reduced, but none want to bear the costs of reducing their own emissions. Though these models have been invaluable in helping scholars understand the initial challenges in the give-and-take of emission control, Cameron's speech suggests that broadening the perspective can identify areas that could be win-win. We analyze this win-win narrative.

Cameron focuses on issues such as fossil fuel subsidies, high-volume horizontal hydrofracturing (fracking), and free trade. Rather than attempting to fit this speech into a conventional "collective action problem" narrative—that is, rather than taking an existing model and adding more assumptions to it until it generates speeches like this as part of its predictions—we can begin a new narrative. Thus, we use this story as inspiration for a different angle on international climate change negotiations. Our query becomes **Query 1**: Why would the British prime minister seek to lower the standards of a climate change treaty?

To address this query, we begin by revisiting some important parts of the story.

Narrative 6.1: Agenda-Setting to Get Consensus in Paris

Prime Minister David Cameron has called for the international community to come together behind a "new, ambitious global deal" to restrict global warming to 2C. But the PM insisted that there was no need to choose between economic growth and reducing carbon emissions, telling the UN Climate Summit that innovative technologies—including shale gas fracking and nuclear energy—could help rein in climate change.

. . . Mr Cameron . . . made clear that any deal in Paris must involve legally binding commitments for all countries around the world.

"We now need the whole world to step up to deliver a new, ambitious, global deal which keeps the two-degree goal within reach," Mr Cameron told the summit. "And I will be pushing European Union leaders to come to Paris with an offer to cut emissions by at least 40% by 2030."

Mr Cameron called for leaders to work hard to "raise the level of ambition" at Paris and prevent a repeat of the failed Copenhagen summit in 2009, warning: "We cannot put this off any longer. To achieve the deal we need all countries to make commitments to reduce emissions," he said. "Our agreement has to be legally binding, with proper rules and targets to hold each other to account. And we must provide support to those who need it, particularly the poorest and most vulnerable."

Mr Cameron said it was "completely unrealistic" to expect poorer countries to forgo the carbon-fuelled economic growth enjoyed by the West. But he insisted: "If we get this right there's no need for a trade-off between economic growth and reducing carbon emissions. . . . We need to give business the certainty it needs to invest in low carbon."

"That means fighting against the economically and environmentally perverse fossil-fuel subsidies which distort free markets and rip off taxpayers. It means championing green trade, slashing tariffs on things like solar panels. It means giving business the flexibility to pick the right technologies for their needs."

"In short we need a framework built on green growth not green tape."

He added: "As political leaders, we have a duty to think long term. . . . And when presented with an opportunity to safeguard the long term future of our planet and our people, we should seize it. I would implore everybody to seize this opportunity over the coming year.

. . . Friends of the Earth's Campaigns and Policy Director Craig Bennett . . . said: "Arriving at a climate change summit with a speech that promotes fracking is like trying to sell cigarettes at a hospital."

. . . Greenpeace UK climate campaigner Sara Ayech said: "David Cameron was right to go to the summit and back a binding global deal . . ."

That a story in this chapter leads to a single narrative does not reflect a scarcity of potential narratives. Indeed, as the chapter unfolds, we refer to the possibility of many others. We limit ourselves to one narrative in order to spotlight the comparison of different methodological approaches to modeling the same narrative. Instead of adopting alternative theoretical angles across narratives, we now alternate methods across models. As we alternate methods, we with readers shift gears between first thinking about the preference orderings of actors, and then considering their absolute utility gains instead.

6.2. Social Choice Agenda-Setting Model

Note that by raising issues such as fracking, free trade, and business investment in a speech on climate change, the prime minister draws the ire of some environmental groups. They accuse him of lowering the standards for a treaty. This sharpens the puzzle: Why is Cameron bringing in these features? What are the items of contention that the prime minister juggles, as he manipulates the leaders of the global community in pursuit of his goals?

Six distinct features seem to characterize the possible future international regime. These features appear disjointed, with some more relevant to some nations than others, and yet these are the building blocks that Cameron seeks to exploit in his "packaging" exercise. To clarify, Cameron engages in a process of agenda setting that entails linking together different issues: bundling items in policy packages, so as to present potential signatories with the composite bundles rather than with individual items when they have choices to make at the drafting stage. In rearranging the set of alternatives in this way, Cameron aims to gain consensus so widespread that it attains the level of unanimity among the participants in support of one of his more preferred alternatives versus one of his lower-ranked preferences.

Specifically, in order to have a *universal* treaty signed, each signatory would have to prefer whatever package the treaty contains to the status quo. Thus, at the last stage, the goal of the agenda setter, shared by all participants designing the content of the treaty, is to have the treaty as a package unanimously preferred to the status quo. Before that is put to the test, however, the content of the agreement must be hammered out. The decision rule on what the package up for signing into a treaty could include is less demanding; for instance, we can approximate it as some form of a majority rule for modeling purposes.

Notably, Cameron does not present the document as being endlessly customizable. Instead, Cameron's agenda-setting consists of limiting and structuring the set of alternatives that can be considered as the treaty's content: instead of letting participants pick or discard items at will, he aims to force participants to compare full documents. That is, the bundles of items that he assembles are to become the only alternatives he intends to have on the agenda for the leaders of the global community.

The narrative identifies the basic features that the prime minister is juggling as follows.

Features:

A global deal in Paris with a unanimous and legally binding commitment to reduce carbon emissions enough to limit global warming to 2 degrees Celsius	*Paris*
Members of the EU commit to cut emissions by at least 40 percent by 2030	*EU*
Permit cheap extraction methods for fossil fuel, e.g., fracking, and permit nuclear power	*Fracking*
Ban fossil fuel subsidies (which "distort free markets and rip off taxpayers")	*Ban subsidies*
Free trade in "green" goods and technology	*Free trade*
A trade-off between economic growth and reducing carbon emissions, resulting in breaks for Non-Annex I (NAI) countries	*Breaks for NAI*

As seen in the narrative, Cameron intentionally avoids raising the possibility of directly comparing these discrete outcomes and avoids, too, making a determination on each in isolation from the others. Instead, as emphasized, he "packages" them together, presenting the international community with outcome bundles, suggesting that the potential signatories to the future (Paris) climate agreement should be asked to make choices among these bundles. Indeed, he almost implies that these bundles of his creation are the *only* feasible alternatives that can be compared and agreed upon. If he were to succeed, in the sense that other international leaders also came to view all these issues as linked, he would reset the agenda for the treaty process in the direction that would manifest the consensus required for passage—and the sort of consensus achieved would be one that he himself could accept. As detailed below, the decision rule for signing the treaty is unanimity, and the technical agenda entails an up or down vote on whatever package the treaty document encapsulates.

We now more specifically label as *packaged alternatives* the bundles composed of the features listed above. As the news story conveys, Cameron in his speech and as the agenda setter works his way methodically through a number of different alternatives that the international community could choose. But, unlike a good social choice theorist who would enumerate the complete set of *all* possible alternatives, Cameron directs attention to a small, restricted set of prepackaged alternatives, as follows.[3]

Packaged alternatives

Paris + EU + free trade + breaks for NAI + fracking + ban subsidies
(Proposal)
Paris + EU + free trade + breaks for NAI + fracking + subsidies
(Proposal + subsidies)
Paris + EU + free trade + breaks for NAI + ban fracking + subsidies
(Proposal – fracking)
– Paris – EU – free trade + breaks for NAI + fracking + subsidies
(no Paris)
– Paris – EU + free trade + breaks for NAI + fracking + subsidies
(no Paris, but free trade)

Let us review the content of these packages that Cameron implicitly invites global leaders to evaluate and compare. To conduct this review, we first need to identify to whom Cameron appeals. Then we turn to appraising these actors' preference orderings over Cameron-defined packages of features.

Actors

Prime Minister Cameron   (C)
Signatories to Paris who are fracking stakeholders   (US)
Signatories to Paris who are EU members   (EU)
Signatories to Paris from Non-Annex I countries   (NAI)

Starting with the prime minister himself, the first alternative on the list is his actual proposal for a climate change treaty, to limit temperature increases to 2° C, which he advances in his speech. This does not necessarily mean that there is nothing in the world that he would like more, but this may be the top strategically attainable priority. Thus, there may be alternatives on the list that he personally would rate higher. Supposing that we can take him at face value, and that attaining *Paris* is of the highest importance to him, anything with *Paris*, he should rank higher than anything with *no Paris*. Furthermore, we can accept from his speech that he wants to see green outcomes, and so *free (green) trade* should be better for him than *no free trade*, other things equal. Cameron also seems vehement on the subject of fossil consumer subsidies, so we can rank an alternative with *ban subsidies* above that with *subsidies*, as long as everything else is the same. Observe that he also appears to dangle the promise of pushing for an even more stringent environmental regulation for Europe—contingent on the success of the universal agreement. The prime minister's preference ordering then looks as follows:

PM preference ordering

Paris + EU + free trade + breaks for NAI + fracking + ban subsidies
(Proposal)

Paris + EU + free trade + breaks for NAI + fracking + subsidies
(Proposal + subsidies)

Paris + EU + free trade + breaks for NAI + ban fracking + subsidies
(Proposal – fracking)

– Paris – EU + free trade + breaks for NAI + fracking + subsidies
(no Paris, but free trade)

– Paris – EU – free trade + breaks for NAI + fracking + subsidies
(no Paris)

Alternatively, we can represent Cameron's preferences as

Proposal $>_C$ *Proposal + subsidies* $>_C$ *Proposal + ban fracking* $>_C$ *no Paris, but free trade* $>_C$ *no Paris*

where " $>_C$ " can be read ". . . is strictly preferred by Cameron to . . ."

But let us now revisit why we assume that unanimity is needed for treaty adoption. Any sustainable climate change agreement needs widespread endorsement. Previous climate change mitigation treaties, such as the 1998

Kyoto Protocol, incorporated a mechanism whereby the treaty would only take effect once a certain fraction k of the n signatories had actually ratified the treaty. The working assumption was that international leaders would become signatories, and then later would undertake and complete their own domestic ratification process (e.g., Munasinghe and Swart 2005). Specifically, in the Kyoto Protocol, the treaty would take effect once it had been ratified by countries that represented at least 55 percent of the total emissions across the developed world (Vanderheiden 2009). For current purposes, we can simplify this to $k = 0.55$. The choice of k, however, is a problem that lurks beneath the surface of Cameron's speech. Although the Kyoto Protocol did eventually reach its 55 percent ratification threshold, the process took years, and several countries—notably, the United States—refused to ratify. The Kyoto Protocol was always weakened by this lack of participation. In practice, it was impossible for a decision to become internationally binding for any $k < n$. Hence the quest for unanimity in 2014.

We return to Cameron, then. The prime minister needs cooperation from the other potential participants in the Paris Treaty: Signatories to Paris who are fracking stakeholders (US), Signatories to Paris who are EU members (EU), and Signatories to Paris from Non-Annex I countries (NAI). Consider their preferences in that order. Widespread fracking in the United States by September 2014 and Americans' general addiction to cheap energy potentially mean that fracking could be the pivot for the US for participation in the global treaty, if banning fracking were on the agenda for a binding climate agreement. Note that fracking cannot be banned without such an agreement, since in that case national laws would apply, and so the combination of "no Paris" and "no fracking" is not practicable. Thus, for the United States, anything with *fracking* is preferred to anything with *ban fracking*. Beyond that, if there were an agreement to be concluded, the United States would prefer to participate in it. At the time of the 2014 story, the United States likes *EU* and *free trade*, since the first improves the global environment at no additional cost to the United States, and the second applies to fields where American technology might be at the forefront. As analysts, we have discretion in deciding on the assumptions to adopt about the US preference with regard to *breaks for NAI*. On the one hand, this does not serve to increase the competitiveness of American products. On the other hand, without *Paris*, there is no issue of whether there would be *breaks for NAI* granted to non-Annex I economies: those economies would be left free to pollute according to their national laws. We choose to assume that the United States is indifferent where it comes to that outcome. Similarly, we choose to assume that the United

States is indifferent on the issue of *subsidies*. This reasoning brings us to order alternatives for the United States as follows.

US preference ordering

Paris + EU + free trade + breaks for NAI + fracking + ban subsidies
  (Proposal)
Paris + EU + free trade + breaks for NAI + fracking + subsidies
  (Proposal + subsidies)
− Paris − EU + free trade + breaks for NAI + fracking + subsidies
  (no Paris, but free trade)
− Paris − EU − free trade + breaks for NAI + fracking + subsidies
  (no Paris)
Paris + EU + free trade + breaks for NAI + ban fracking + subsidies
  (Proposal − fracking)

To summarize, we can represent the US preferences as

$$Proposal \succeq_{US} Proposal + subsidies >_{US} no\ Paris,\ but\ free\ trade >_{US} no\ Paris$$
$$>_{US} Proposal + ban\ fracking$$

where "$>_{US}$" can be read ". . . is strictly preferred by the US to . . ." and "$\succeq$ US" can be read ". . . is viewed by the $_{US}$ with indifference compared to . . ."[4]

Next, we will assume that Britain's compatriots in 2014—EU members—prefer any form of a global agreement over *no Paris*. We will assume, however, that they also want *no fracking* and *subsidies*. We can justify their preference for *ban fracking* not only by direct environmental concern about that technology but also by the reduction in the cost of fossil fuel associated with fracking, which makes green energy less attractive. As the EU members have a technological advantage in green energy, they should like *free trade*, and we can assume that they are indifferent with regard to *breaks for NAI*. Their preference ordering is then:

EU preference ordering

Paris + EU + free trade + breaks for NAI + ban fracking + subsidies
  (Proposal − fracking)
Paris + EU + free trade + breaks for NAI + fracking + subsidies
  (Proposal + subsidies)

Paris + EU + free trade + breaks for NAI + fracking + ban subsidies
(Proposal)
– Paris – EU + free trade + breaks for NAI + fracking + subsidies
(no Paris, but free trade)
– Paris – EU – free trade + breaks for NAI + fracking + subsidies
(no Paris)

Or, expressed as a preference relation,

Proposal – fracking $>_{EU}$ *Proposal + subsidies* $>_{EU}$ *Proposal* $>_{EU}$ *no Paris,*
but free trade $>_{EU}$ *no Paris*

where "$>_{EU}$" can be read ". . . is strictly preferred by EU to . . ."

The final actor whose preferences Cameron must anticipate in his agenda-setting is the set of signatories to the Paris Treaty from Non-Annex I countries, NAI. They cannot afford to sign any agreement that does not include *breaks for NAI*, as we just heard Cameron state.[5] But observe that this is already incorporated in the alternatives that Cameron offered. His agenda setting is about which possibilities to consider, and he rules out including a proposal for Paris without *breaks for NAI*. Meanwhile, if the agreement fails, then developing countries do not have to restrict their carbon emissions at all. The outcome of *breaks for NAI* or better is thus attained regardless of which of his alternatives does prevail. This actor, however, is interested in having others sign the agreement (prefers *Paris*), and in having EU members subject themselves to even more stringent environmental regulations (prefers *EU*), because both lower the competitiveness of industries in developed countries: at least in the short run, both amount to an increase in manufacturing and agricultural costs for developed countries. We further assume that NAI are indifferent on *free trade* insofar as green technology is concerned, and very slightly prefer not to *ban fracking*, and prefer *subsidies* when it comes to domestic fuel consumption. This leaves us with the following preference ordering.

NAI preference ordering

Paris + EU + free trade + breaks for NAI + fracking + subsidies
(Proposal + subsidies)
Paris + EU + free trade + breaks for NAI + ban fracking + subsidies
(Proposal – fracking)
Paris + EU + free trade + breaks for NAI + fracking + ban subsidies
(Proposal)

– Paris – EU + free trade + breaks for NAI + fracking + subsidies
(no Paris, but free trade)
– Paris – EU – free trade + breaks for NAI + fracking + subsidies
(no Paris)

Or, expressed as a preference relation,

Proposal + subsidies $>_{NAI}$ *Proposal – fracking* $>_{NAI}$ *Proposal* $>_{NAI}$ *no Paris,
but free trade* $\simeq_{NAI}$ *no Paris*

where "$>_{NAI}$" can be read ". . . is strictly preferred by NAI to . . ." and
"$\simeq_{NAI}$" can be read ". . . is viewed by NAI with indifference compared
to . . ."

In the social choice situation that we now find ourselves, thanks to
Prime Minister Cameron, with four actors and five alternatives, let us calculate which alternative will emerge as a winner in some sort of decision
process, where decision makers understand that they will require unanimous consent to whatever they would put on a final, yes or no, agenda.

Normally, to do so, we would need to know the voting weights of each
actor, because most social choice situations call for decisions by some sort
of a majority. Once again, however, the goal of a universal and binding
agreement implies that the *final decision rule* in this case is by *unanimity*. We
assume that decision making in earlier stages, where alternatives are compared by the international leaders as they formulate the terms of the treaty,
proceeds by majority rule, with two out of three "voting" actors—US, EU,
and NAI—sufficient to make a choice.

Table 6.1 represents the preference profile for the three types of voting actors in Model 6.1. In turn, Model 6.1 in figure 6.1 depicts a binary
agenda that pairs all alternatives against each other as well as against the
status quo at the final, unanimity stage. Observe that, because *no Paris +
free trade* is ranked the same or immediately above *no Paris* by all players
in table 6.1 and strictly preferred by the agenda setter, we only include *no
Paris + free trade* in the agenda tree. As pictured, first Cameron's Proposal
is "voted" on against the same proposal but without the ban on fossil fuel
domestic subsidies (*Proposal + subsidies*). Next the winner is paired with the
same proposal, but which also bans fracking (*Proposal – fracking*). The winner of that "ballot" is then compared with the alternative of not signing the
Paris Treaty but formalizing the free green trade agreement (*No Paris + free
green trade*). In reality, all these "ballots" are decided by some potentially
semiformal process that we model here as majority rule among the three

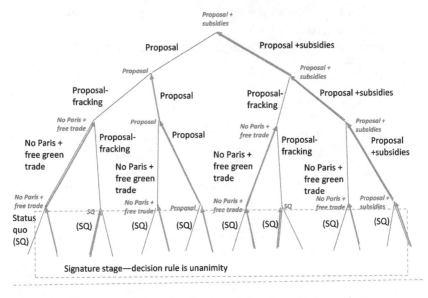

Figure 6.1. Binary agenda based on Cameron's "packaged" alternatives

TABLE 6.1. The preference profile of actors in the global decision-making body

Cameron	US	EU	NAI
Proposal	Proposal; Proposal + subsidies	Proposal–fracking	Proposal + subsidies
Proposal + subsidies	No Paris, but free trade	Proposal + subsidies	Proposal–fracking; Proposal
Proposal–fracking	No Paris	Proposal	No Paris, but free trade No Paris
No Paris, but free trade	Proposal–fracking	No Paris, but free trade	
No Paris		No Paris	

"actors" (actually, groups of actors). At the final ballot, the decision rule changes from majority to unanimity, so that the surviving alternative, now the document remaining on the table, must be endorsed by all in order to win, or else the reversion alternative, the status quo, will prevail.

Agenda trees are analyzed from the top down when actors are believed to be voting their sincere preferences on each ballot. They are instead analyzed by backward induction, from the bottom up, when we have reason to expect actors to anticipate what would happen on subsequent ballots and take these consequences into consideration whenever they are called to vote. Since we, as well as Prime Minister Cameron, can rightly believe that all actors in this particular social choice model are *sophisticated*, that is, that when voting on alternatives they consider and anticipate the prospective outcomes obtainable down the road, rather than the immediate choices with which they are presented, we analyze the agenda from the last ballot and move from the bottom up.

All actors know that, after they draft the document, by successively comparing the alternatives as defined by Cameron, the alternative they select will then "run" against the status quo by the unanimity rule at the signing stage. We also assume that all actors know each other's preferences, just as they know the sequence of ballots (the agenda) that leads to the final outcome. Because of that, as the actors make their choices within pairs of alternatives, they base their vote on what they want the final outcome to be, and not on their true preferences between the currently compared alternatives. Figure 6.1 represents this in two ways. First, it traces the path of anticipated voting decisions with arrows, starting from the last ballot, which in our case, again, is by unanimity. Second, in accordance with the arrows, it reports *strategic equivalents* of voting for each alternative in bold italics at the bottom of a corresponding branch except for the last ballot. Together, these two features capture the process of roll back followed by a sophisticated voter: at each ballot, she votes for her preferred strategic equivalent associated with an actual ballot alternative.

Once we have put together figure 6.1, it solves our social choice model. Observe that on the last ballot sophisticated actors vote sincerely because, at that point, they get exactly what they voted for as the outcome. Since the outcome is the status quo unless all support the alternative, *status quo* becomes the strategic equivalent of voting for *Proposal – fracking* on the penultimate ballot: the United States prefers *status quo* to *Proposal – fracking*, so unanimity is out of reach when it comes to signing that version. Every other alternative beats the status quo by unanimity, becoming strategic equivalents of voting for those alternatives on the previous ballot. Moving one step up, note that the strategic equivalents of voting for *Proposal – fracking* on the second ballot are *No Paris + free green trade*.

On the first ballot, then, we truly compare *Proposal* with *Proposal + subsidies*, as these are the same as their strategic equivalents. We assume the United States to be indifferent, and the EU members to prefer to keep the subsidies. If we were now to question the latter assumption, supposing that Britain's leadership could sway the collective preference in Europe against the subsidies, then the first ballot would become a tie. Either way, Cameron's agenda-setting effort, if implemented, assures him of a smooth and seemingly consensus-based adoption of one of his top two alternatives. Considering that a strong consensus within the international community is also one of his stated goals, his agenda-setting effort can be deemed a success.

The model we have designed also vindicates Cameron in his defense of fracking practices. It looks like sophisticated actors who anticipate that the final selection would have to be approved by unanimity would, indeed, fail to draft the final agreement that would include any ban on fracking, because the United States would prefer to block the deal. Hence the analysis shows that Cameron's agenda setting would accomplish his primary goal—his move to not ban fracking would prevent American withdrawal. But his secondary goal of banning fossil fuel subsidies likely would not be served, at least not if actors' preferences are as we assumed them and the issue of including subsidies is open for preliminary debate.

6.3. The Same Narrative Modeled as a Cooperative Game

In our social choice model for the narrative, we have included actors' preferences without accounting for the relative intensity of those preferences— without asking "how much" actors like or dislike the alternatives in relation to each other. Indeed, with regard to some outcomes, actors could have a preference, but be almost in a take-it-or-leave-it state. Sometimes this is due to true indifference, but sometimes this reflects ambivalence instead, as there are too many conflicting implications and it is hard to estimate with accuracy whether gains would outweigh the costs. Other items, like a fracking ban for the United States, feature as an unambiguous deal breaker for that actor in the model. Now we repeat the same process under the same general assumptions as before, but give actors cardinal utilities, so that we can compare the magnitude of their gains or losses from each alternative with regard to the status quo. We continue presuming that Cameron has a goal of building a coalition for the unanimous support of the ultimate treaty package. We move, in other words, to design a cooperative game-theoretic model for this narrative.

As seen in table 3.1 in chapter 3, cooperative game theory deals with *outcomes* (what in section 6.2 we called *alternatives*), and judges each outcome in terms of whether or not it can beat the status quo by gaining the unanimous preference of at least one decisive coalition. In our story, that translates to finding alternatives that can attract unanimous support. In a cooperative game-theoretic model, Cameron's agenda manipulation in the narrative would entail placing an alternative in the policy space in such a way as to unsettle the status quo. Cameron's packaging of items into bundles amounts in effect to constructing *multidimensional* alternatives, that is, the policy space in which he battles the status quo is multidimensional.

The main difficulty with this specific narrative is that we have multiple actors, and each actor is in competition with each other actor at least on the economic dimension, but different pairs of actors compete over different items. Because of this, the trade-offs that they face would need to be considered in the context specific to each pair: EU-NAI, EU-US, and US-NAI. For example, a bundle that Cameron proposes might improve the position of the EU relative to NAI countries, but worsen the position of the EU relative to the United States. The sufficient condition for a successful policy proposal to defeat the status quo by unanimity under such circumstances is that *every* competitive pair would obtain "gains from trade" from the proposal relative to the status quo. Consider what would happen otherwise: if an actor were to benefit vis-à-vis one "trading partner" but lose in a relationship with the other partner, we as modelers (or Cameron as a heresthetician) would have no means of numerically comparing the gain and the loss without very precise information about the utility function of the actor in question. This is why we choose to avoid this additional complication and instead pursue the *sufficient* condition of benefitting each actor in *each* of its pairwise relationships for the unanimous approval of the proposed treaty.

Consider first that the policy space where the prime minister is aiming to defeat the current policy status quo has a dimensionality at most equal to the number of features that he packages in his alternatives (bundles) as in section 6.2. This is because the largest number of dimensions that may exist are the yes or no extremes regarding each feature separately. That, however, is likely to overestimate the dimensionality of the space that Cameron constructs, because some of the items can be placed plausibly at different points on the same dimension or a few dimensions. We can start constructing the space in which "exchange" contained in the drafting of a mutually beneficial agreement is taking place by checking the dimensions on which various features can be compared. As we do so, we also seek to identify whether their values are greater or lower for actors relative to the status quo state.

Take the pair of actors Countries of the European Union and NAI Countries (figure 6.2). Recall that the items in use by Cameron are as fol-

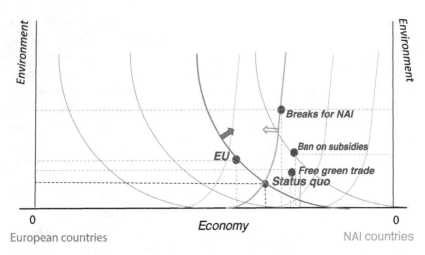

Figure 6.2. Items under consideration in the speech, positioned in terms of gains from trade between countries of the EU and NAI countries

lows: *EU, fracking, subsidies, free trade, breaks for NAI*. Considering these in turn, *EU* (an EU-wide policy that would reduce at once the pollution production and relative competitiveness of EU industries, a promise that Cameron dangles before the world leaders and that is unrealized at the moment of delivering his speech) is not a part of the status quo. Can other items be placed on the same economic-environmental continuum with *EU?* The first thing that comes to mind is *breaks for NAI*, since reducing the competitiveness of European industries equals more *breaks for NAI*, at least insofar as the dimension of economic competitiveness is concerned. Meanwhile, the status quo in the pre-treaty world should be economically beneficial compared to *breaks for NAI*, as far as Non-Annex I countries are concerned vis-à-vis the countries of Europe—considering that no "breaks" are required before having imposed any binding constraints. The status quo also includes less environmental protection, and we can safely assume that the countries of Europe as well as Non-Annex I countries at least weakly prefer a better environment to a worse environment.

Figure 6.2 presents the Pareto-improving set in the two-dimensional economy-environment space for the actors. The representation is in the form of the Edgeworth box, where the gains for the bargaining sides are increasing in opposite directions on the economy.[6] The policy space for the countries of Europe is constructed at one origin in bottom left of the figure, and the reciprocal space for Non-Annex I countries spans out from the other origin, at the bottom right, with the economic axis now going from right to left. Since both value the environment and gain on the environment simultaneously, the environment axis is aligned between them.

Other items on the list can also be projected on the same two-dimensional plane at least for the two actors in figure 6.2. Thus, *fracking* and *subsidies* are both a part of the status quo. Both actors would gain on the environment if fracking were banned, but since they do not trade with each other on the economic axis in connection with fracking, it is not an issue in their pairwise economic bargaining and so does not appear in figure 6.2.

Despite the prime minister's assertions, the economic model rationale forces us to admit that *free green trade* is an economically better proposition for some than it is for others. Having the dimension of the economy separate from that of the environment, we can resolve the ambiguity from the previous section and assume that NAI countries bear economic losses from *free green trade* vis-à-vis the EU. Specifically, tax-free global sales of green technologies benefit the exporting countries and deprive the governments of importing countries of their fiscal revenue. We can thus compare *free green trade* with the status quo in figure 6.2 as follows: it is better for the European actor and worse for the Non-Annex I actor on the economic dimension, while better for both on the environmental dimension.

Note that we assume in drawing figure 6.2 that the individual items *EU* and *Breaks for NAI* are on the same indifference curves for EU and NAI, respectively, as is the status quo. Assuming for simplicity that the effects of (i.e., gains or losses from) the items are additive on each of the two dimensions, consider now figure 6.3, representing the trade-offs between the same two actors, but now showing the locations of Cameron's "packaged" outcomes that he puts forward. Notice that gains from *breaks for NAI* would partially offset losses from *EU* for Countries of EU on the economy, and the reverse applies for NAI countries. Meanwhile, an additive improvement on the environment accrues to both actors. The combined package of *EU* and *Paris with breaks for NAI* thus is located in the Pareto-improving set—in the mutually beneficial set—for these two actors.

The *Countries of NAI* lose economically if *free green trade* is added to the package (*EU + Paris + breaks for NAI + free green trade*). If both *free green trade* and *ban fracking* are added, NAI in figure 6.3 prefers to keep the status quo, as *EU + Paris + breaks for NAI + ban fracking + free green trade* is at a lower indifference level for them than the status quo (per assumptions that we made when placing items in the issue space). Note that *ban subsidies* hurts NAI on the economy the most—enough to offset both the environmental and economic benefits from *EU* and *Paris + breaks for NAI*, so any package that includes *ban subsidies* is outside the Pareto lens in figure 6.3.

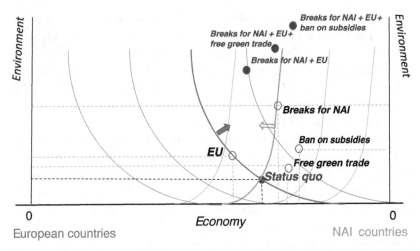

Figure 6.3. "Gains from trade" between the countries of the EU and NAI countries

Aside from banning fossil fuel subsidies for domestic consumers, which seems to be a deal breaker in figures 6.2 and 6.3, packaging multiple features into the bundled alternatives allows Cameron to place his resulting alternatives in the Pareto-improving set of the two actors. Insofar as Countries of EU and NAI countries are concerned, that pair would consider as mutually beneficial relative to the status quo both *EU + Paris + breaks for NAI + ban fracking* or *EU + Paris + breaks for NAI + free green trade*, despite their dislike of *free green trade* and a possible slight dislike of *ban fracking*. This, however, does not guarantee the support of the two actors for either proposal. In order to get that support, each of them must not lose in their relationship with the third actor, the United States, relative to the status quo.

Since the trade-offs differ when it comes to respective gains or losses within different pairs of actors, we now repeat the analysis for US and NAI. Consider figure 6.4. Even though both these actors would reap overall economic benefits from unilateral European limits on emissions (*EU*), in this they do not benefit at each other's expense, and figure 6.4 reflects that fact by placing *EU* higher on the environmental dimension for both, but at the same location as the status quo on the economic dimension.

The two features, ban on subsidies and free green trade, appear in similar relative locations to those in figure 6.3. US gains and NAI countries lose on free green trade on the economic dimension. A ban on fracking is mutually disliked, though its placement in figure 6.4 reflects our assumption that it is disliked more by the United States and less by NAI countries.

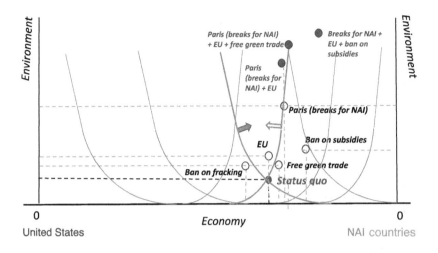

Figure 6.4. "Gains from trade" between the United States and NAI

On balance, we give NAI the economic advantage over the US from adding that feature to *Paris*.

Prime Minister Cameron thus can plausibly believe that *EU + Paris + breaks for NAI, EU + Paris + breaks for NAI + free green trade, EU + Paris + breaks for NAI + free green trade + ban fracking* and even *EU + Paris + breaks for NAI + ban subsidies + ban fracking* are all located in the US-NAI Pareto lens, and are thus mutually beneficial outcomes for those two actors.

Finally, for Countries of EU and US, figure 6.5 provides similar analysis. No economic shift between these two actors is assumed to take place from either green free trade or a ban on subsidies. Both would benefit from those on the environmental dimension, and so packages that include these features would merely move up on the vertical dimension compared to the packages without. A ban on fracking puts the US at a relative economic disadvantage. So does participation in the Paris Treaty, exacerbated by a slower increase in US utility (steeper indifference curves) with environmental improvement, when compared to that of Countries of EU. Unilateral European limits on emissions (*EU*) sufficiently offset *Paris* for the US, considering that it also improves the environment, but it cannot do so for both *Paris* and *ban fracking*. The Pareto set of Countries of EU and US then contains the alternatives *EU + Paris, EU + Paris + free green trade, EU + Paris + ban subsidies*, and *EU + Paris + free green trade + ban subsidies*.

Thus, every competitive pair as captured in figures 6.2 through 6.5 is

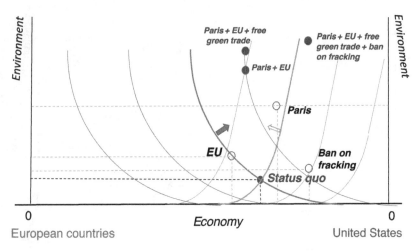

Figure 6.5. "Gains from trade" between the countries of the EU and the US

appeased by Cameron's agenda setting via packaging of the items. There are now bundles of his creation on the proverbial table that he can have unanimously supported by the international community over the status quo. The alternatives on Cameron's list that are mutually beneficial for all three actors and that thus satisfy the sufficient condition for beating the status quo by unanimity rule are *EU + Paris (breaks for NAI)* and *EU + Paris (breaks for NAI) + free green trade.*

6.4. No Testable Propositions: Outcomes of Agenda Manipulation versus Counterfactuals

What we have done in this chapter does not exhaust the analytical possibilities. We could position our query to apply other modeling methodologies. Or we could pose other queries. We could start with Prime Minister Cameron himself as a single actor. We might think about why this speech was given, using a decision-theoretic framework to analyze why Cameron chose to give this speech rather than a different speech—or chose to give a speech in the first place. This could be a productive exercise for those interested in executive politics: Why do chief executives make the decisions they do?

We might also think about consequences of the speech. Once it was made, what happened? What were the consequences for Cameron? What were the consequences for any other actors that we might want to bring into the analysis? These research questions would move us outside the decision-theoretic framework, and into a game-theoretic framework. We would be focusing on the implications of that speech for others—perhaps other international leaders, or perhaps a domestic audience. These narratives lead us out of executive decision making and into climate change treaty negotiations (e.g., Barrett 1999; Hovi et al. 2015; Madani 2013), into signaling to domestic political audiences (e.g., Putnam 1988; Tingley and Tomz 2014) and to subnational governments (e.g., Bechtel and Urpelainen 2015), or into the relationships between politicians and interest groups (e.g., Grossman and Helpman 2001).

That said, what we did in this narrative closely resembles what students of analytic narratives do in their analyses: we sought an explanation (in our case, a rationalization) of an observed social phenomenon. Because the phenomenon is unique, generalization to a broader class of observations is not immediately apparent, and testable propositions are not forthcoming.

6.5. Conclusion

In this chapter, we have seen that different assumptions as dictated by different methods yield somewhat different findings. We have seen as well that we as analysts have no a priori or even a posteriori way to judge one method as superior to another. In using both social choice and cooperative game theoretic models, we make and also trace through the assumptions that are explicit and thus can be altered if we disagree with them or find that they are not useful. Doing so would be analogous to the process of reconciliation as outlined in chapter 3.

SEVEN

Games and Uncertainty in U.S. Criminal Justice Systems

With Andrei Zhirnov, *Binghamton University*

This chapter offers alternative narratives derived from a news story about abuse of power in the Sheriff's Department of Los Angeles County. The story poses multiple puzzles, each of which features a set of actors engaged in strategic interactions. The narratives we construct from this story and the models we distill from them form the core of this chapter. All models in this chapter illustrate the use of noncooperative game theory.

This chapter stands as our last application of the algorithm for narratives and models. We deliberately place this chapter, with its noncooperative game theoretic models, after chapters 4 through 6, which have illustrated, in sequence, utility-maximization, decision-theoretic, social choice, and cooperative models. We proceed in this way because, as noted, the components of a noncooperative game theoretic model constitute a superset of the first three models just named, and contrast well with the components of a cooperative model. Alongside the diversity of methods across chapters is the commonality of epistemology that they all share and that is first captured in figure 2.1.

Story:

Six L.A. County Sheriff Workers Get Prison for Obstructing Jail Probe

Victoria Kim

Los Angeles Times | *2014-09-23*

A federal judge on Tuesday lambasted what he called a "corrupt culture" within the Los Angeles County Sheriff's Department as he sentenced six current and former members of the department to prison for obstructing a federal investigation into abuse and corruption at the county jails.

U.S. District Judge Percy Anderson said evidence showed there were "significant problems" within the department, including an "us-versus-them mentality," routine cover-up of inmate abuse, and an "unwritten code" taught to new jail deputies that any inmate who fought a guard should end up in the hospital.

While the six defendants were not accused of excessive force, they were complicit in such misconduct by trying to thwart a federal investigation into abuses at the jails, Anderson said. Jurors convicted the sheriff's officials this year of conspiring to impede a grand jury investigation by keeping an inmate informant hidden from his FBI handlers, dissuading witnesses from cooperating and trying to intimidate a federal agent.

"They all took actions to shield these dirty deputies from facing the consequence of their crimes," Anderson said.

Sentencing the sheriff's officials to terms ranging from 21 months for one deputy to 41 months for a seasoned lieutenant, the judge castigated the defendants for not showing "even the slightest remorse." He said he intended for the sentences to send a message that "blind obedience to a corrupt culture has serious consequences."

"The court hopes that if and when other deputies face the decision, they will remember what happened here today . . . and they will do what is right rather than what is easy," Anderson said.

Interim Sheriff John Scott said in a statement that the punishments "reflect individual action by a few."

"Those sentences should not be seen as a broad brush characterization of the quality of work performed, or commitment to the public, the men and women of the Los Angeles County Sheriff's Department deliver each and every day," he said.

The five men and one woman stood largely expressionless as they were sentenced.

Their attorneys took turns pleading for leniency, citing their long histories of service as sworn law enforcement officers. They blamed higher-level officials in the department, echoing arguments made at trial that the defendants had merely followed orders from above.

Peter Johnson, an attorney for Lt. Stephen Leavins, asked the judge to take into account the role then-Sheriff Lee Baca and Undersheriff Paul Tanaka played in "orchestrating and guiding" actions taken by the subordinates.

© InlandPolitics.com

This story about the Los Angeles County Sheriff's Department is but a speck in an avalanche of problems in the U.S. criminal justice systems. Those who are incarcerated in the United States have the legal right to humane treatment, with federal agencies designed and presumed to ensure enforcement of that right. A body of social science research investigates the actual treatment of the incarcerated in the United States, the unusually high U.S. incarceration rates, when viewed in cross-national perspective, and the dramatic racial disparities in U.S. incarceration rates (e.g., Gottschalk 2008; Lerman and Weaver 2014; Weaver and Lerman 2010). To illustrate, over one-tenth of African American men aged twenty-five to twenty-nine were incarcerated as of 2007, and one third of African American men aged twenty to twenty-nine were under some sort of correctional supervision as of 2008 (Weaver and Lerman 2010, 817).

Given the conditions in the contemporary U.S. criminal justice systems, the term "carceral state" has become commonplace. Prisoners' human right to humane treatment is all too often observed in the breach in the United States, as nongovernmental organizations, advocacy groups, and media observers have established (e.g., American Civil Liberties Union 2011; Amnesty International 2014; Fellner 2015; Southern Poverty Law Center 2017; Woolf 2016). Social scientists working in different disciplines come to the same conclusion (e.g., Bayley 2002; Gottschalk 2008; Weaver 2007; Weaver and Lerman 2010). In the United States, documented cases of maltreatment of prisoners amounting to torture, and documented cases of police brutality, including killings, arise with alarming frequency.

For the American public, any news of abuse in the U.S. criminal justice systems acquires salience and resonance, and can galvanize countermobilization based on clashing views and interpretations. Further confounding the public, the plural we just used reflects the coexistence of diverse, overlapping, and often disjoint systems for criminal justice under U.S. federalism, a trait recognized by scholars and advocacy groups alike (e.g., National Center for Victims of Crime 2019; Petersilia 2014; Wilson and Petersilia 2011; Young and Petersilia 2016). This multiplicity of systems recurs as a feature of our analysis here.

Given this context, what question will the modeler choose to investigate when presented with a real-world set of events, as in this particular story of abuse? Once again, the scholar's choice of question drives the content of the narrative that she draws from a news story. For instance, suppose first that the analyst's pressing question is what the incentives might be to motivate the repeated, systematic abuse of prisoners carried out by the guards. That abuse is not only widespread but also assumes the repetition, routine, established expectations, and clout of an institutionalized practice. Put simply, why would—why did—the guards work so hard to abuse the prisoners?

7.1. Why Do Guards Abuse Prisoners?

Our first query thus becomes: *Query 1*, Why do guards abuse prisoners? Why did the deputies act the way they did, abusing prisoners and imposing the same behavior on their new colleagues? How can their actions be explained given the high potential costs to them if prosecuted? Why did all these people display little remorse over their behavior? Observe that as we move through the query, its parts, and its accompanying narrative, we identify and develop its theoretical implications in multiple literatures.

7.1.1. Rationalizing Norms of Illegal Behavior within Law Enforcement

Narrative 1 distills the original story to focus on the events and facts needed to address the theoretical implications of *Query 1*. The focus in this narrative is on the pervasiveness and uniformity of the illegal behavior among the deputies, and their observed belligerence in defending such illegal behavior. The overall inference in the narrative—its punch line—is that, however illegal and detestable the deputies' actions might appear to someone else, the deputies' actions were, at least to them, a necessary answer to some larger problem that they all as a group shared and that they each individually faced.

The inquiry into the conditions under which actors endowed with power abuse their power—and into the institutions that might curb abuses—is central to political science. This news story also calls up such classic questions as how multiple layers of institutions interact with one another, mesh, or fail to do so; how socially shared expectations become established, are maintained, or on the other hand erode or crumble; and, more broadly, what factors drive institutional change.

Our focus in this case is on the deputies: what they did, how uniform their participation in the illegal practice was, and whether they seem to have shared a belief that their actions were somehow justified or even just. Actions and statements by other potential actors from the original story are not relevant for answering **Query 1**, because other actors in the story arrive on the scene *after* the pattern of behavior persists, and persists for a long while. The other actors take the stage too as a consequence of having discovered such patterns of behavior. We implicitly group all those additional actors and their actions as described in the story as part of those "high costs" that transgressing deputies know they might face but that fail to stop them from committing the transgressions. With that in mind, the story's essentials can now be framed in narrative 7.1. Once again, we see the distinction between the actors who the theorist views as relevant for the strategic equilibrium and those who do not play a role in this particular strategic outcome. The distinction and its significance are flagged in the way the narrative 7.1 unfolds.

Narrative 7.1: Why Do the Deputies Feel Justified in Adopting Their "Unwritten Code" of Inmate Abuse?

A federal judge on Tuesday lambasted what he called a "corrupt culture" within the Los Angeles County Sheriff's Department as he sentenced six current and former members of the department to prison for obstructing a federal investigation into abuse and corruption at the county jails.

U.S. District Judge Percy Anderson said evidence showed there were "significant problems" within the department, including an "us-versus-them mentality," routine cover-up of inmate abuse, and an "unwritten code" taught to new jail deputies that any inmate who fought a guard should end up in the hospital.

While the six defendants were not accused of excessive force, they were complicit in such misconduct by trying to thwart a federal investigation into abuses at the jails, Anderson said. Jurors convicted the sheriff's officials this year of conspiring to impede a grand jury investigation by keeping an inmate informant hidden from his FBI handlers, dissuading witnesses from cooperating and trying to intimidate a federal agent.

"They all took actions to shield these dirty deputies from facing the consequence of their crimes," Anderson said.

. . . The judge castigated the defendants for not showing "even the slightest remorse."

. . . The five men and one woman stood largely expressionless as they were sentenced.

This narrative implies the existence of a shared goal resulting from some shared threat. It also implies a high level of consensus and commitment among the transgressing deputies as well as among their supervising officers in the Sheriff's Department. The researcher thus advances the theoretically based proposition that some other actor or actors are the protagonists of the story whose importance is equal to that of the deputies, and who are essential to understanding the deputies' behavior. These other actors pose a threat to each and every one of the deputies that is sufficient to motivate all individually, from supervisors to new recruits, to use force. Such a threat, moreover, leads them to deploy force at a level exceeding that required for suppressing individual acts of dissent, and leads them as well to consider their participation in illegal abusive behavior, if not just, then justified and even inevitable. Such a level of threat suggests that the shared protagonists are the inmate population: some high level of threat emanates from them.

7.1.2. The Game between Guards and Prisoners

The simplest modeling choice is to model all deputies as one, and all prisoners as one, since the narrative and the theoretical proposition suggest consensus within each group. Thus, we start with two players, Guards and Prisoners, who are playing the game where prisoners pose some level of threat, and Deputies can act to deter or not, but the deterrence option comes with some personal costs for the Deputies. There can be a number of versions of such a game. We prefer the one we present given its advantages of intuitive appeal, simplicity, and power.

As shown in figure 7.1.1, Prisoners get to choose among three types of action: they can keep quiet; they can engage in low-level, expressive dissent; or they can rebel to take over the prison. If Prisoners keep quiet, both players get the payoff of 0, as indicated at the terminal node on the top of figure 7.1.1. If Prisoners choose the expressive form of dissent, they receive an expressive benefit b, regardless of Guards' reaction. (Table 7.1.1 records this and all other notation used in figure 7.1.1.) Guards can either give a strong, h, or a weak, l, response to the disobedience, $r \in \{h,l\}$. Focus for

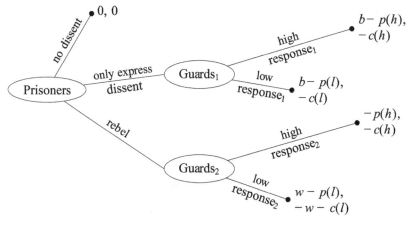

Figure 7.1.1. A game between prisoners and guards

now on the Prisoners' option of rebellion. If the Guards choose a strong response, they bear the cost of $c(h)$ and impose the cost (punishment) of $p(h)$ on the Prisoners, as represented in the payoffs at the corresponding terminal nodes on the right side of Figure 7.1.1. If the Guards choose a weak response, they bear the cost of $c(l)$ and impose the cost of $p(l)$ on the Prisoners. If the strength of reaction is insufficient to suppress a rebellion (i.e., it is weak), rebellion succeeds. The very fact of rebellion can cost any guard her life, thus giving her a very low payoff. A successful rebellion costs more lives in the Deputy population, and so costs an individual deputy her life with a higher probability, bringing her payoff further down, as represented at the bottom terminal node of the figure, with the notation $-w-c(l)$ for the Deputies. A successful rebellion also transfers the value of w from the Deputies to dissenters, that is, to the protagonists who have attained the status and power of Rebels. The Rebels' payoff thus appears at the bot-

TABLE 7.1.1. Notation for figure 7.1.1

Symbol	Meaning
b	expressive value of dissent
w	value of the control over prison, $w>b$
$r \in \{h, l\}$	intensity of response, high and low
$c(r)$	costs borne by guards as they engage in repression at level r, $c(l)<c(h)$
$p(r)$	costs imposed by guards on prisoners as they engage in repression at level r, $p(l)<p(h)$
s	organizational sanction

tom terminal node as $w-p(l)$: they pay $p(l)$ to mount the rebellion but they wrest away from the Guards the value of dominance, of winning, that is, w.

In this game, we assume that the Guards observe the level of dissent before choosing their response. This characterizes the situation when their reaction is not institutionalized.[1] To give a prediction for this game, we apply the solution concept of the subgame perfect Nash equilibrium to the game in figure 7.1.1. Under this solution concept, a combination of strategies is considered a solution if it constitutes an equilibrium in each of the proper subgames of the game. This solution concept guarantees that, each time the players make a decision, they act in accordance with all the information they have at their disposal; this has become a standard tool in complete information games with more than one proper subgame. Such solutions are typically found via backwards induction: the analyst goes through each decision-making contingency (information set), reversing the order of play, that is, starting from the terminal nodes and finishing at the root, eliminating all actions other than best responses at each step.

Table 7.1.2 lists the equilibria of this game. If $w<p(l)$, the game is trivial: the Prisoners do not dissent because they lack sufficient motivation. In other cases, whether the Prisoners engage in dissent and what type of dissent they choose depend on the Guards' valuation of control over the prison relative to the additional costs of strong retaliation and, accordingly, on the Guards' expected reaction to dissent. If $w>c(h)-c(l)$, Guards would choose a strong reaction to rebellion, thus inducing Prisoners to avoid rebellion and engage only in expressive dissent. If $w<c(h)-c(l)$, Guards would not dare to give a strong reaction to strong dissent. It is precisely this—possible—structure of payoffs that motivates the Prisoners to rebel. And it is here that the need arises for a strong norm among the Guards: if the choice of reaction were left to the discretion of a specific Guard, and this Guard did not have immediate incentives to issue a *high response*, his response would jeopardize the safety of the entire prison.

Recalling the scenario of Rebellion equips the analyst to take up the situation in which Guards commit to retaliate with full strength to any form of dissent, be it violent or merely voiced. The Guards' choice to institutionalize the strong response to dissent can be represented as an additional stage in the game. The Guards' choice to institutionalize strong retaliation precedes and must be assumed to structure the Guards' payoffs, as represented in figure 7.1.2. In subgame B, the Guards' payoffs decrease by s—the sanction implied in the "unwritten code" and imposed by the community and organization—whenever they fail to react to the dissent in full strength.

Figure 7.1.2 shows this compound game in the extensive form. Subgame A is the case in which the Guards choose not to institutionalize a high level of retaliation. Subgame B is when Guards choose to institutionalize high retaliation. To find subgame perfect Nash equilibria of this game, we examine subgames A and B separately. Subgame A reproduces the game presented in figure 7.1.1 (with some superficial adjustments); thus, its behavioral predictions are the same as in that earlier game.

The predictions of subgame B, however, differ. When the Guards, on the contrary, choose to institutionalize a *high response* policy by setting the sanction to $s>c(h)-c(l)$, that is, high enough to force any guard with any valuation w of control over the prison to retaliate with full strength, then the Prisoners will not rebel. If the Prisoners do rebel, they will not and cannot succeed: this combination of strategies brings the cost of $p(h)$ to them. In addition, the Prisoners will engage in expressive dissent only if $b>p(h)$, which is a rather demanding condition. Except for these extreme circumstances, neither actor suffers any losses. If the Prisoners do opt for expressive dissent, the Prisoners receive $b-p(h)$, and the Guards pay the cost of $c(h)$.

Table 7.1.3 shows the conditional values of subgames A and B to the Guards in this compound game. A comparison between them reveals the conditions when the institutionalization of a high retaliation policy makes sense to the Guards. If $p(h) > b > p(l)$, the Guards are clearly better off institutionalizing the high retaliation policy. Only under the extreme conditions of $b > p(h)$ are the Guards better off not institutionalizing the high retaliation policy.[2]

The game as set up and analyzed in figure 7.1.2 reveals why and how in equilibrium illegal abuse becomes routine, expected, and institutionalized. This institutionalized abuse means that Rebellion, in equilibrium, will

TABLE 7.1.2. Subgame perfect equilibria

Condition 1	Condition 2	Prisoners' strategy	Guards' strategy	Payoff to Guards	Payoff to Prisoners
$w < p(l)$	$w > c(h) - c(l)$	no dissent	Low response$_1$, high response$_2$	0	0
$w < p(l)$	$w < c(h) - c(l)$	no dissent	Low response$_1$, low response$_2$	0	0
$b < p(l) < w$	$w > c(h) - c(l)$	no dissent	Low response$_1$, high response$_2$	0	0
$b < p(l) < w$	$w < c(h) - c(l)$	rebel	Low response$_1$, low response$_2$	$-w - c(l)$	$w - p(l)$
$b > p(l)$	$w > c(h) - c(l)$	express only	Low response$_1$, high response$_2$	$-c(l)$	$b - p(l)$
$b > p(l)$	$w < c(h) - c(l)$	rebel	Low response$_1$, low response$_2$	$-w - c(l)$	$w - p(l)$

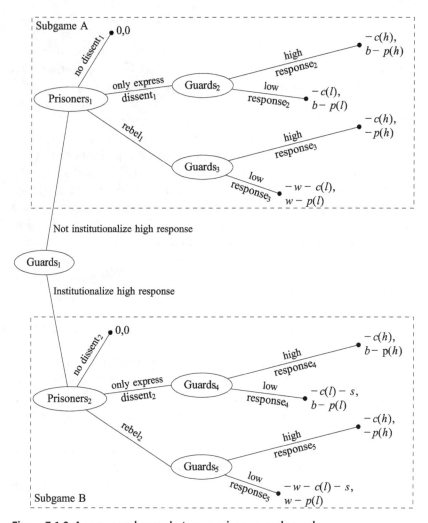

Figure 7.1.2. A compound game between prisoners and guards

TABLE 7.1.3. The conditional values to the guards of proper subgames A and B of the compound game

Condition 1	Condition 2	Subgame A		Subgame B
$w < p(l)$	$w > c(h) - c(l)$	0	$=$	0
$w < p(l)$	$w < c(h) - c(l)$	0	$=$	0
$b < p(l) < w$	$w > c(h) - c(l)$	0	$=$	0
$b < p(l) < w$	$w < c(h) - c(l)$	$-w - c(l)$	$<$	0
$p(h) > b > p(l)$	$w > c(h) - c(l)$	$-c(l)$	$<$	0
$p(h) > b > p(l)$	$w < c(h) - c(l)$	$-w - c(l)$	$<$	0
$b > p(h)$	$w > c(h) - c(l)$	$-c(l)$	$>$	$-c(h)$
$b > p(h)$	$w < c(h) - c(l)$	$-w - c(l)$	$>$	$-c(h)$

not be played by the Prisoners. The Guards' commitment to a strong and indiscriminate retaliation to any dissent serves as a necessary and sufficient deterrent to inmate dissent. If such a reaction is not institutionalized, the Guards are likely to face frequent acts of small-scale dissent and, under certain conditions, rebellions.

7.1.3. Hypotheses from Model 7.1

If we were now inclined to move beyond explaining the event and toward generalizing patterns of behavior under similar circumstances, then we would now turn, as we do, to generating hypotheses based on model 7.1. In doing so, we discover similarities with the conjectures in a body of work on criminal justice. Our conclusions in model 7.1 closely resemble recurring findings in a prominent school of research on corrections (e.g., O'Hara 2005; Silberman 1995). According to this school, violence forms part of the system of norms internalized by both guards and inmates. As O'Hara (2005, 149) puts it, "The members of an institutionalized agency believe firmly in the goodness of their processes, their goals, their colleagues and their right to be largely free from outside intrusion." In this account, just as it happens to be in our model, when the inmates are ready to use violence at any moment, the use of violence by correctional officers becomes necessary not only as a means of checking outbreaks of dissent and extinguishing internal disputes but also as a tool for maintaining authority. Indeed, in the strategic logic in model 7.1, the avoidance of violence is viewed as a sign of weakness and an invitation to dissent. Wilsnack (1976) documents that, under conditions of organizational weakness and high rates of turnover among correctional personnel, the correctional officers cannot internalize norms and are unable "to respond effectively to strains on inmates before the strains result in a riot" (1976, 69). As in our model, the appropriate hypothesis for Wilsnack (1976) is that such weakness, turnover, and inability to internalize collective norms on the part of correctional officers should be good predictors of inmate rioting.

Such a logic is familiar to scholars of international relations. The idea of massive retaliation has long figured prominently in the deterrence literature (e.g., Powell 1990, 1999; Schelling 1960, 1966). This model is also comparable to the logic presented in Selten's (1978) Chain Store Paradox. In that problem, managers of a chain store have to make decisions about their response to the entry of other firms in the markets where the chain has its stores. The paradox postulates that the chain store managers might be better off adopting a rule of a strong reaction to any entry, which acts as a deterrent against any and all competitors. Unlike our model, Selten's (1978) analysis focuses on the idea that such a strategy is not dynamically consistent: once the competitor has entered, managers do not have enough motivation to fulfill their threat. The chain store paradox is typically resolved through the notion of reputation. In repeated interactions, that is, the managers of the chain store benefit from waging a price war against the entrant as it signals the seriousness of their threat against any future entrant. In our example, the indiscriminate retaliation rule is institutionalized as a part of the informal code, which makes it self-enforceable in the short run.

The discussion thus far points to several hypotheses. The greater the frequency of such observable signs of organizational weakness in prison administration as turnover or internal conflicts among personnel, labor-management disputes, or change in management, the weaker should be the adherence to internal norms among the Guards. According to our model, then, the weaker should be the deterrent effect with regard to the Prisoners' behavior. This leads us to three testable propositions.

H7.1.1: The greater the organizational weakness in the prison administration, the more frequent the incidence of violence and unrest among the prisoners should be.

This is a direct implication of the organization's inability to impose a sufficiently high sanction, s, for the failure to retaliate.

H7.1.2: In prison communities with a relatively recent experience of expressive dissent, the norm of illegal behavior on the part of the guards should develop relatively quickly. The more recent the experience of inmate dissent, the more rapidly the norm of abusive behavior among the guards should spread and should become entrenched.

This indicates that condition $p(h) > b > p(l)$, which is most conducive to the high retaliation policy, is satisfied.

H7.1.3: In prison communities with relatively recent instances of guards' failure to suppress extreme violence (primarily in the absence of intermediate levels of dissent), the norm of illegal behavior on the part of the guards should develop relatively quickly.

This indicates that condition $w < c(h) - c(l)$, which is most conducive to the high retaliation policy, is satisfied.

The abuse is now comprehensible: it emerges from model 7.1.2 as an equilibrium institution for running the prison.

7.2. Why Do the Deputies Obstruct the FBI Investigation?

The Guards continued to perpetrate and perpetuate the abuses while knowing that Federal Agents were conducting an investigation. Even more than that, the Guards obstructed the federal investigation. That fact leads to our second query. The second narrative and the model we develop from it center on the puzzles inherent in the interactions between the deputies and the FBI. Why did the FBI conduct the investigation covertly, using an informant, and not overtly? What, if anything, did the deputies gain in the end by behaving in such a perplexing way? Rational choice theory requires that the observed behavior be rationalizable—whatever the actual outcome, was it even possible for the deputies to benefit from it? These questions culminate in **Query 2**: What benefit did the Sheriff's deputies hope to obtain by obstructing the federal investigation and interfering with officials they knew to be covert operatives?

7.2.1. Risk, Obstruction, and Monitoring

The narrative here focuses solely on the puzzle of local law enforcement officers impeding the ability of agents of another branch and layer of law enforcement to contact their associates with information from their investigation. The observer knows little with certitude other than the fact, affirmed in court, that the obstruction took place. We also have information on the steep cost to the deputies who chose to become involved in obstructing the federal investigation. Narrative 7.2, then, focuses on these aspects and dismisses everything else as irrelevant. As before, we narrow our attention to the elements that illuminate the chosen query and promise to offer the greatest analytical leverage on it. We again winnow the story's details to construct the narrative, and then abstract away from the narrative to build the model.

Narrative 7.2: What Hope of Success Can the Deputies Have in Obstructing a Federal Investigation?

A federal judge on Tuesday lambasted what he called a "corrupt culture" within the Los Angeles County Sheriff's Department as he sentenced six current and former members of the department to prison for obstructing a federal investigation into abuse and corruption at the county jails.

. . . While the six defendants were not accused of excessive force, they were complicit in such misconduct by trying to thwart a federal investigation into abuses at the jails, [U.S. District Judge Percy] Anderson said. Jurors convicted the sheriff's officials this year of conspiring to impede a grand jury investigation by keeping an inmate informant hidden from his FBI handlers, dissuading witnesses from cooperating and trying to intimidate a federal agent.

"They all took actions to shield these dirty deputies from facing the consequence of their crimes," Anderson said.

. . . Their attorneys . . . blamed higher-level officials in the department, echoing arguments made at trial that the defendants had merely followed orders from above.

Peter Johnson, an attorney for Lt. Stephen Leavins, asked the judge to take into account the role then-Sheriff Lee Baca and Undersheriff Paul Tanaka played in "orchestrating and guiding" actions taken by the subordinates.

This narrative names a number of aspects of bad behavior that the deputies shared in addition to abusing the prisoners. The deputies conspired to impede a grand jury investigation, they deliberately isolated the inmate informant from the FBI investigators, and they pressured and intimidated potential witnesses. The deputies went so far as to try to intimidate a federal agent. Moreover, their attorneys subsequently argued that the fault for any and all of this behavior did not lie with the deputies, and that they should not have been held accountable for any and all of it: the deputies "had merely followed orders" from their superiors.

Hence the narrative—even as it conveys the sentence for obstruction of a federal investigation—paints elusive and slippery behavior on the part of the deputies, behavior that obscures the issue of their guilt. This point merits emphasis: the deputies as *agents* to the public—the *principal*—strive to render their actions unobservable and unverifiable. They strive to obscure their behavior to the federal officials as the principal's representative in this narrative. If the federal officials investigate, they run the risk of turning up nothing; plainly, an investigation that yields no results wastes the costs of monitoring. The issue for them becomes: Is an investigation worthwhile? We refer to the principal in a theoretical sense, as an actor who is monitoring and attempting to ascertain the true type of the agent, again in a theoretical sense. In an administrative sense, there is no superior-subordinate relationship between these branches of law enforcement (e.g., FBI 2017). With the theoretical conceptualization of agents and principals firmly in mind, we bring to light the logic underlying the deputies' choices, made despite a steep cost attached to obstruction, by designing a new game that focuses now on interactions between the deputies and the federal agents.

7.2.2. The Game between the Sheriff's Deputies
and the Federal Agents

In this game, the players are the Federal Agents and the Sheriff's Deputies. The Federal Agents suspect inmate abuse in the prison and launch an investigation. They know that a Sheriff's Department can be of two types: an organization with a strong unwritten code—in the sense described in section 7.1, that is, the one that can impose behavioral norms—and an organization with a weak unwritten code (types A and B, respectively). The Deputies' type is their private information, which is not available to the Federal Agents. At the start of the game, the Federal Agents believe that the probability that the Sheriff's Department under investigation is of type A, that is, organizationally strong, is π_0.

The Deputies, who know the organizational strength of their department, choose between obstructing the investigation and cooperating with Federal Agents. If the Deputies choose to cooperate, the game is over, and the Federal Agents receive reward r for the conviction of the guilty. The payoff to the Deputies depends on the organizational strength of the Sheriff's Department: if it is strong, in addition to the punishment p for the inmate abuse, they receive the organizational punishment s for cooperation with another law-enforcement agency. Indeed, for an organization with a strong code, an act of covert investigation—that is, an investigation that bypasses the leadership of that organization—should be felt as a gross violation of that organization's autonomy, which must be resisted by all means. Cooperation with such an intruder is a punishable offence for the members of such an organization.

If the Deputies choose to obstruct the investigation, the Federal Agents may either persist or give up. If the Federal Agents give up, the game is over, and both players receive 0. If the Federal Agents persist, the Deputies are punished both for the inmate abuse and the obstruction of justice (their payoff is $-p-q$), and the Federal Agents are rewarded for conviction of the guilty parties. In this case, the payoffs obtained by Federal Agents also depend on the organizational strength of the Sheriff's Department—if it is strong, it can impose cost c on the persistent Federal Agents, leaving them with the payoff of $r-c$. The c term is multifaceted: it can be interpreted as a decrease in the probability of unearthing evidence and convicting the guilty, as well as the material costs and risks to the personnel involved in pursuing such evidence.

To derive the predictions for this game, we employ the solution concept of a perfect Bayesian equilibrium (PBE). A combination of strate-

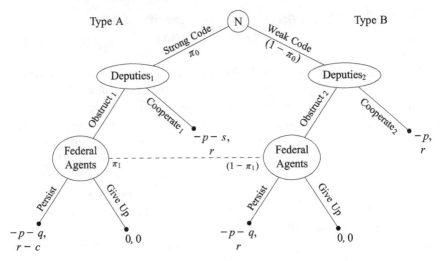

Figure 7.2 A game between deputies and federal agents

gies, together with players' beliefs at each decision-making contingency, is considered a perfect Bayesian equilibrium if and only if the choices they suggest to a player at each of his decision-making contingencies give that player the highest expected utility given that player's beliefs at that time. Also, all players' beliefs must be consistent with their initial beliefs and the play they observe (and follow Bayes' rule). This solution concept is widely used in incomplete-information games and is somewhat analogous to that of a subgame perfect equilibrium.[3]

Table 7.2.2 lists the perfect Bayesian equilibria of the game. If $s<q$, that is, if the costs that a Sheriff's Department with a strong unwritten code can impose on its members are lower than the costs of judicial punishment, the game is trivial: Sheriff's Departments of both types are expected to behave the same way, and the play provides no additional information to the Fed-

TABLE 7.2.1. Notation for figure 7.2.1

Symbol	Meaning
π_0, π_1	probability that the informal organization of deputies is strong
p	punishment for inmate abuse
q	punishment for obstruction of justice
s	organizational sanction
r	reward for finding the abusing party
c	cost imposed by organizational resistance to the investigation

eral Agents. Having identified this equilibrium as a relatively trivial case, we set it aside.

If the organizational sanctions, s, that a strong Sheriff's Department can impose on its members are on the contrary relatively high, then the outcome depends on the Federal Agents' beliefs about the strength of the Sheriff's Department and their expectations with respect to the Deputies' guilt and, accordingly, their expected reward for cracking the case. When $r>c$, the Federal Agents are unstoppable—they will persist in their investigation no matter what. In this case, the Deputies from a department with weak norms will have no incentives to obstruct justice, while the Deputies from a department with strong norms will be forced to obstruct anyway. Thanks to the different expected behaviors of the Deputies of different types, the observation of Deputies' actions will reveal their type to the Federal Agents: if the Federal Agents see obstruction, they can be certain that the Sheriff's Department has a strong unwritten code. Such an equilibrium is called a separating one.[4]

If $r<\pi_0 c$, that is, the expected reward for pursuing the investigation is low, and the Federal Agents have incentives to give up on the investigation. In this case, the Deputies always obstruct the investigation, even if they are not pressed by their unwritten code. In this case, too, the Federal Agents cannot update their beliefs by observing the Deputies' actions.

If the size of the reward is moderate, $\pi_0 c<r<c$, we do not have a perfect Bayesian equilibrium in pure strategies. We do have one in mixed strategies, as now seen. Deputies from a department with a strong informal code choose to obstruct the investigation, while Deputies from a department with a weak informal code mix between obstructing the investigation (with probability $(\pi_0 c - \pi_0 r)/(r - \pi_0 r)$) and cooperating with the Federal Agents (with probability $(r - \pi_0 c)/(r - \pi_0 r)$). The Federal Agents persist with probability $p/(p+q)$ and give up with probability $q/(p+q)$. They cannot completely determine the type of Deputies and believe that the Sheriff's Department has a strong code with probability r/c.

TABLE 7.2.2. Perfect Bayesian equilibria in figure 7.2.1

	Conditions		Deputies' strategy	FAs' strategy	FAs' Beliefs π_1
1	$s > q$	$\pi_0 c > r$	obstruct $_1$, obstruct $_2$	give up	π_0
2	$s > q$	$\pi_0 c < r < c$	obstruct $_1$, obstruct$_2$ with probability $\pi_0(c - r)/(1 - \pi_0)/r$	persist with probability $p/(p + q)$	r/c
3	$s > q$	$r > c$	obstruct $_1$, cooperate $_2$	persist	1
4	$s < q$	$\pi_0 c > r$	obstruct $_1$, obstruct $_2$	give up	π_0
5	$s < q$	$\pi_1 c < r$	cooperate $_1$, cooperate $_2$	persist	exogenous

This result is somewhat counterintuitive. To appreciate this point, consider what would be a complete information counterpart of this game. Suppose that the Federal Agents knew which type of Deputies they faced, that is, instead of having one information set, the Federal Agents had two information sets, one for each state of the world. We already know what happens if $r>c$: in this case, the predictions will be the same as in the separating equilibrium treated above: the Federal Agents are undeterrable and unstoppable. If $r<c$, the Federal Agents facing a Sheriff's Department with a strong code have incentives to give up, while the Federal Agents facing a Sheriff's Department with a weaker code have incentives to persist with their investigation. Expecting this play, the Deputies from a strong department will obstruct the investigation, and the Deputies from a weak department will cooperate. In such a modified game, the obstruction is a means whereby the Deputies from an organizationally strong department communicate a threat in order to deter the Federal Agents from pursuing their investigation. Arguably, this is the motivation behind the obstruction of justice presented in the narrative.

In the original game, however, the only circumstance when the Deputies can communicate their organizational strength is when the Federal Agents cannot be deterred. A game with less motivated Federal Agents does not produce a separating equilibrium because the Deputies from organizationally weaker departments have incentives to mimic the behavior of Deputies from type A departments. Given such incentives, it takes higher expected costs of obstruction, c, to deter the Federal Agents from pursuing their investigation, as transpires from a comparison between rows 1 and 4, on the one hand, and rows 2 and 3, on the other hand, of table 7.2.2.

The narrative that drives this model is consistent with observing the realization of the mixed strategy equilibrium (see row 2 of table 7.2.2). In this equilibrium, Deputies of both types engage in obstruction of the investigation, although the type B Deputies also cooperate with a positive probability. The FBI abandons its investigation with probability $q/(p+q)$, which increases with the relative scale of punishment for the obstruction of justice vis-à-vis the punishment of inmate abuse. Since the existence of this equilibrium is contingent on the model's parameters meeting the specific conditions (in row 2 of table 7.2.2), we could check the validity of the model by examining the relevant characteristics of the penitentiary system and of the specific prison.

7.2.3. Hypotheses from Model 7.2

We now turn to extracting hypotheses based on model 7.2, recognizing that the conclusions of model 7.2 align with salient themes in extant scholarship. A number of studies (e.g., Brehm and Gates 1999; Sieberg 2005) view police brutality and excessive use of force as one of the outcomes of principal-agent problems in law enforcement. The everyday risks involved in police work not only can justify the use of violence by police officers but also can provide the excuse for unjustified, excessive use of force. Moreover, the institutionalization of practices in law enforcement agencies makes it almost impossible to detect malpractices and generate honest self-reporting.

While in-person monitoring might be useful in providing information on officers' actions, such activities might be associated with prohibitive costs (e.g., Brehm and Gates 1999; Kiewiet and McCubbins 1991; Kornblum 1976). According to Sieberg (2005), assigning an actor such as a federal agent to monitor police officers is the costliest type of monitoring, and also exposes the monitors to the same risks as police officers.

The point of model 7.2 is that a stronger informal code in a Sheriff's Department and stronger norms of organizational autonomy do not necessarily bring more safety from monitoring to the deputies. Even though an agent of a particular type (to remind, a Deputy) can impose prohibitively high costs of monitoring, such costs do not always deter the principal (here, the Federal Agent, acting for the public). As shown, the obstruction can probabilistically fail to convince Federal Agents to back off in their monitoring efforts; yet in the face of considerable resistance, Federal Agents might still persevere.

The key to this result is the uncertainty surrounding the intent and ability of the deputies to conceal evidence and otherwise hinder the investigation. This is not the only type of uncertainty in the story. Obstruction here is aimed at preserving the principals' uncertainty about the agents' guilt. The agents use obstruction to sow confusion within the organization as well as to increase the costs of outside monitoring. The agents, by engaging in ongoing obstruction, perpetuate incompleteness of information, by preventing principals from updating their beliefs. It is possible to approach the principals' problem from another direction: another way of saying that obstruction imposes prohibitively high costs of effective monitoring is to say that obstruction makes it harder (rather than impossible) for the principals to update their beliefs. Such an alternative conceptualization would lead to a model with a different state space and different payoffs. In the current model, the term r represents the likely reward for a successful investigation and, by extension, encapsulates the players' expectations about the existence of evidence that could prove the deputies' guilt. In an alternative model with Nature choosing the extent of the deputies' guilt, the value of r would be contingent on the state of the world.

These two sides of the same theoretical coin lead to different operationalizations for empirical assessment captured in the following hypotheses.

H7.2.1: If investigators think that deputies are guilty with very high probability, only departments with very strong informal codes will try to obstruct an investigation.

These are the conditions for the separating equilibrium of this game: such an equilibrium is realized only if the expected rewards for persisting with the investigation, r, are high.

H7.2.2: The lower the expected award for persisting with an ongoing investigation, the higher the probability that investigators will drop their inquiries, holding constant the burden of proof (parameter c), the extent of the "normal" use of extreme force in the prison industry (π_0 and π_1), and the strength of the sanctions imposed by the informal code relative to the strength of legal sanctions (s and q).

Five parameters—s, q, r, c, and π_0—jointly determine the expected behavior of the Federal Agents. Holding constant the other parameters, a higher expected award for persisting with the investigation (parameter r) is associated with equilibria in which Federal Agents drop their investigation with lower probability. The expected award for persisting with the investigation includes the rewards offered by the FBI leadership, their moral rewards for fighting crime, as well as the chances that "the fish is in the pond," that is, that the crimes have actually been committed.

H7.2.3: The higher the "normal" level of the use of extreme force in the prison industry (π_0 and π_1), the lower the probability of continued investigation for the same values of the other parameters.

Lower values of π_0 are associated with equilibria in which Federal Agents are more likely to persist in their investigations. Since the Federal Agents cannot know for sure whether the department under investigation can persevere in obstructing their activities, they are bound to rely, at least partially, on their prior beliefs about the type of department. Such prior beliefs could come from two sources: ecological knowledge—knowledge about the penitentiary system in general—as well as from the evidence in place by the time of the investigation.

7.3. What Does the Judge Hope to Accomplish beyond Punishing the Guilty?

Now suppose that we shift the focus to an actor in the story that we have so far neglected, the Judge. To analyze the Judge's behavior, we return to the story and extract an alternative narrative. We, then, with the new narrative, build yet another model.

7.3.1. Guilt, Punishment, and Precedent

The Judge in the story seems to have an agenda for the long term that he associates with the punishment that he metes out to guilty Deputies. How can we capture or rationalize a reasonable long-term agenda for him? Specifically, we pose **Query 3**: How would the Judge treat punishment as affecting incentives for future behavior on the part of the police?

Narrative 7.3: What Motivates the Judge beyond Immediate Punishment of the Guilty?

A federal judge on Tuesday lambasted what he called a "corrupt culture" within the Los Angeles County Sheriff's Department as he sentenced six current and former members of the department to prison for obstructing a federal investigation into abuse and corruption at the county jails.

. . . Sentencing the sheriff's officials to terms ranging from 21 months for one deputy to 41 months for a seasoned lieutenant, the judge castigated the defendants for not showing "even the slightest remorse." He said he intended for the sentences to send a message that "blind obedience to a corrupt culture has serious consequences."

"The court hopes that if and when other deputies face the decision, they will remember what happened here today . . . and they will do what is right rather than what is easy," [U.S. District Judge Percy] Anderson said.

. . . The . . . attorneys [for the defendants] took turns pleading for leniency, citing their long histories of service as sworn law enforcement officers. They blamed higher-level officials in the department, echoing arguments made at trial that the defendants had merely followed orders from above.

In this narrative extracted from the original story, we focus on the connection between the level of guilt and the punishment the court metes out. The Judge uses his judicial discretion to raise the level of punishment with the aim of deterring similar future behaviors under similar circumstances, thus dismantling the institutional equilibrium uncovered in model 7.1. The attorneys are disappointed that the mitigating circumstances do not move the Judge to moderate the sentences. The Judge, for his part, rationalizes the severity of his approach with the argument that future Deputies will be better motivated to stand up to all sorts of pressure to break the rules, even pressure that comes down the chain of command. Thus, the Judge uses his ruling to send a signal.

7.3.2. The Game between the Sheriff's Deputies and the Judge

We frame this narrative and this game in the principal-agent theoretical context. We see the Judge as the principal, and the Deputies as his agents. The Judge is certain about the existence of inmate abuse in the prison, yet faces uncertainty about the level of guilt of each individual defendant. The Judge chooses to make the punishment automatic when any violation is observed, regardless of whether or not the agent was at fault or the events conspired against the agent (cf. Cingranelli, Fajardo-Heyward, and Filippov 2014).

What follows exemplifies a quite simple case of contract design with hidden actions (e.g., Bolton and Dewatripont 2005; Salanie 1997). The principal (Judge) designs a contract with the agent (Deputies) in such a way as to make the agent share in the exogenous risks (in this case, the need for the use of violence) and invest effort in mitigating those risks. Relevant scholarship emphasizes that, in order for agents to be willing to enter into contracts where they are punished for any negative outcome, be it accidental or attributable to their behavior, the principal must offer agents a high level of compensation. Contracts in which agents are asked to absorb all the risks without that high compensation will be unstable and will not attract high-quality agents. In narrative 7.3, however, the Judge does not issue an extra reward, but rather increases the expected costs to agents: he issues the punishment while ignoring possible mitigating circumstances. In this case, the contract becomes less attractive to all types of agents, but high-quality agents can compensate with their effort to improve outcomes, whereas low-quality agents cannot. The effect of the "revised" contract following the verdict should thus be to discourage the participation of low-quality agents, whereas high-quality agents should still be willing to accept the contract.

Consider the Sheriff's Deputies. A Deputy's type is the type of norm prevailing in her department: she can be from a department in which the

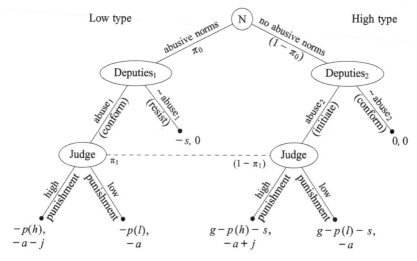

Figure 7.3. A game between the deputies and the judge

leadership and the community instill norms of inmate abuse (low type), or from a department without such norms (high type). The Deputies' type is their private information and is not observable by the Judge. At the onset of the game, the Judge believes that the Sheriff's deputies come from a department with abusive norms with probability π_0. (This and all notation used in figure 7.3.1 appear in table 7.3.1.)

The Sheriff's deputies observe their department's norms and either conform to them or resist. If inmate abuse is the norm, Deputies of low type can choose between obeying and disobeying the norms (information set 1). Absent abusive norms, Deputies of high type can choose on their own initiative to abuse or not to abuse inmates (information set 2).

In the case of low type, the Deputies' resistance is equivalent to staying clean. From the standpoint of those seeking to instill the norm of inmate abuse to deter Prisoners' dissent, a deputy's choice to refrain from excessive violence constitutes an individual defection from the collective efforts meant to create a public good. As earlier, we use the s term to denote the organizational sanction imposed on Deputies who fail to conform to the abusive norms.

In the case of the high type, the choice to resist the norm is the choice to use excessive violence. The choice to resist norms within the second information set is also associated with sanction s. We assume that inmate abuse for a high type Deputy brings her the benefits of g.

Whenever the Judge finds evidence that the Deputies have committed inmate abuse, the Judge incurs the cost of a, the social cost of inmate abuse. He now faces a problem of deciding on the appropriate punishment for the Deputies: whether the punishment will be high or low. In making this decision, the Judge can only see whether the Deputies have committed inmate abuse, not their type. Since he does not know prevailing prison

TABLE 7.3.1. Notation for figure 7.3.1

Symbol	Meaning
$\pi_0, \pi_1,$	probability that the prison has abusive norms
g	personal returns for abusing inmates
$d \in \{l, b\}$	level of punishment
$p(d)$	impact of the punishment for inmate abuse, $p(b) > s > p(l)$
s	organizational sanction
a	social cost of inmate abuse
j	value of justice

TABLE 7.3.2. Perfect Bayesian equilibria in figure 7.3.1

	Conditions	Deputies	Judge	Beliefs, π_1
1	$g > s + p(b)$	$\pi_0 > 1/2$	always abuse	Low punishment	π_0
2	same as above	any π_0	~abuse$_1$, abuse$_2$	High punishment	0
3	same as above	any π_0	abuse$_1$ with probability $\frac{1-\pi_0}{\pi_0}$, abuse$_2$	High punishment with probability $\frac{s-p(l)}{p(h)-p(l)}$	1/2
4	$s + p(b) > g > 2s$	$\pi_0 > 1/2$	always abuse	Low punishment	π_0
5	same as above	any π_0	abuse$_1$ with probability $\frac{1-\pi_0}{\pi_0}$, abuse$_2$	High punishment with probability $\frac{s-p(l)}{p(h)-p(l)}$	1/2
6	same as above	$\pi_1 < 1/2$	never abuse	High punishment	exogenous
7	$2s > g > s + p(l)$	any π_0	abuse$_1$, abuse$_2$ with probability $\frac{\pi_0}{1-\pi_0}$	High punishment with probability $\frac{g-s-p(l)}{p(h)-p(l)}$	1/2
8	same as above	$\pi_0 > 1/2$	always abuse	Low punishment	π_0
9	same as above	$\pi_1 < 1/2$	never abuse	High punishment	exogenous
10	$g < s + p(l)$		abuse$_1$, ~abuse$_2$	Low punishment	1
11	same as above	$\pi_1 < 1/2$	never abuse	High punishment	exogenous

norms, he does not know what share of responsibility for inmate abuse is borne by the Deputies, as shown by the grouping of nodes into the Judge's information sets in figure 7.3.1. Should the Judge decide in favor of high punishment for Deputies, Deputies incur the cost of $p(h)$; if the chosen level of punishment is low, Deputies incur the cost of $p(l)$, $p(l)<p(h)$. We further assume that $p(h) > s > p(l)$, that is, that high punishment suffices to offset the sanctions imposed by the informal norms of the prison, while low punishment does not. The Judge bears the cost j if he assigns high punishment to the Deputies of low type, and gains j if he assigns high punishment to the Deputies of high type. That is, the Judge draws positive utility from giving a severe punishment to the party fully responsible for the observed crime and draws negative utility from giving a severe punishment to those who were forced to commit a crime.

Before proceeding to solve this game for perfect Bayesian equilibria, consider what would happen in a complete information analog of this game. The Judge would give a harsher punishment to the more guilty and a less severe punishment to the less guilty. The Deputies would obey the norm to abuse inmates if such a norm were in place. If the norm were not to abuse (high type), they would defect from that norm if $g > s + p(h)$ and conform otherwise.

Table 7.3.2 lists the cases of perfect Bayesian equilibria of the game in figure 7.3.1. The game has separating equilibria only under two sets of conditions: (1) if the benefit of unilateral inmate abuse is very high, that is, $g > s + ph$, as in row 2 of table 7.3.2; and (2) if the benefit of unilateral inmate abuse is extremely low, that is, $g < s + p(l)$ (row 10). In the first case, the Judge observes inmate abuse only if the prevalent norms prohibit inmate abuse and the Deputies counter them. Thus, he has no reasonable doubt that the Deputies commit inmate abuse of their own will and punishes them to the highest extent whenever he observes such abuse. In the second case (row 10), inmate abuse takes place only if the predominant norms are abusive and the Deputies follow them. In this case, the Judge has no incentives to impose a severe punishment on the Deputies since he can be certain that Deputies simply followed orders.

Under all other conditions, separating equilibria do not exist. When Deputies of both types abuse prisoners, the Judge cannot update his beliefs about norms prevailing in the prison. Thus, if he observes inmate abuse, he cannot determine the Deputies' share of guilt—cannot discern whether circumstances forced them to abuse prisoners or they engaged in abuse on their own initiative. In some equilibrium cases, the update available to the Judge is only partial (this is the case in the mixed strategies equilibria listed in rows 3, 5, and 7).

There are two groups of pooling equilibria in this game: equilibria in which inmate abuse takes place in both states of the world; and equilibria in which the inmate abuse occurs in neither state of the world.[5] In all equilibria belonging to the first group, except for mixed-strategy cases, the Judge cannot update his beliefs and, since all of these cases require that $\pi_0 > 1/2$ and generate beliefs $\pi_1 > 1/2$ at his decision nodes, he chooses low punishment. His beliefs at the nodes in information set 1 are the same as his prior beliefs, that is, that the Deputies come from a department with abusive norms with probability π_0. The second group of pooling equilibria involves Deputies never abusing the inmates, and so the Judge applies severe punishment whenever he observes inmate abuse. In such cases, the existence of equilibria depends on the Judge's off-equilibrium path beliefs—his beliefs on the information sets that should never be reached. These equilibria are played only if the Judge is sufficiently confident that inmate abuse was committed due to Deputies' own initiative (i.e., $\pi_1 < 1/2$).

The Judge's inability to differentiate between innocent and guilty Deputies creates a situation characterized in the contract literature as the moral hazard problem. If the principal expects that under negative conditions agents will perform poorly (here, abuse prisoners when the norms are corrupt) and cannot observe whether conditions in actuality are negative, then agents are tempted to shirk and produce low output even where benevolent conditions prevail.

Notice that there are some combinations of parameters that can lead to multiple equilibria. For instance, if the benefits from unilateral inmate abuse are high $(g > s + p(b))$ and the confidence in the strength of the prison culture is also high $(\pi_0 > 1/2)$, then we can expect a separating equilibrium, a pooling equilibrium in pure strategies, and a pooling equilibrium in mixed strategies. In cases of multiple equilibria, the strategies of a player depend on his or her strategic expectations, that is, the expectations of how other players will behave. In this example, Deputies of high type will not be deterred by the level of punishment. However, if Deputies of low type, the ones from departments with an abusive culture, expect a high level of punishment, they will not transgress; they will on the other hand transgress and follow the norms of inmate abuse if they expect a low level of punishment. One of the interpretations of the Judge's sentence is a coordinating message, which seeks to channel the future play of this game into an equilibrium with relatively low abuse.

We may want to take another step and connect the narrative to the equilibria in which inmate abuse occurs only off-equilibrium path: after all, such equilibria are the most desirable ones for the principal. Suppose that the benefits from unilateral inmate abuse are relatively high, so that $ph + s > g > pl + s$ (rows 6 and 9 in table 7.3.2). The Deputies' strategy is (~$abuse_1$, ~$abuse_2$) and the Judge's strategy is (*high punishment*). This equilibrium further requires that, according to the Judge's off-equilibrium path beliefs (π_1, the beliefs held by the Judge if he observes inmate abuse), the probability that inmate abuse is caused by corrupt norms is less than 0.5. While the scale of punishment can be manipulated by sentencing and by law, the communication of beliefs is trickier and, as it seems in our narrative, can be achieved through announcing the strategy of this clean equilibrium in the cases where the Deputies' actions suggest that they are playing a "dirty" equilibrium with inmate abuse and low punishment (moving the path of play from row 1 to 2, from row 4 to 6, or from row 8 to 9).

This equilibrium is the most desirable one from the standpoint of a Judge: in all other equilibria, inmate abuse actually takes place. The expected value of this equilibrium for Judges is 0, which is the highest for this game for almost any combination of parameters. The expected value of this equilibrium for the Deputies is, however, only $-s\pi_0$.

The negative value of the game discloses features of the Deputies who would self-select into such a contract. Hold constant for a moment the reward Deputies receive for service in Sheriffs' Departments, the prevalence of inmate abuse across departments, and the strength of informal abusive rules. Deputies with a high susceptibility to following the prevailing societal norms on respecting human dignity are less likely to self-select into this contract. This is so because the wage (w_0) must compensate for any losses incurred from combating the departmental norm of inmate abuse, so that the net benefit is $\bar{w} = w_0 - s\pi_0$, and constitutes all the compensation for the labor and the professional risks of a law enforcement officer.

7.3.3. Hypotheses from Model 7.3

This model speaks to major themes in the literature on police brutality. For some time, scholars have studied the general problem of moral hazard in police forces (e.g., Brehm and Gates 1999; Kornblum 1976; Sieberg 2005). The inherent risks in policing as well as the informal rules prevalent in law enforcement agencies, known as police culture or institutionalization of practices (e.g., Brehm and Gates 1999; Crouch and Marquart 1989; O'Hara 2005; Silberman 1995; van Maanen 1983), often yield the result that law enforcement officers use force even if alternative instruments are

available. Since actual decisions on the use of force are taken by police officers at their own discretion, such circumstances and norms can serve as an
excuse for unjustified violence.

Sieberg (2005, 171–80) suggests using the tools of contract design
to mitigate the problem of agents' hidden actions in law enforcement.
Whereas the general contract literature suggests that the principal can
change the agent's incentives by setting output-contingent wages, as this
analysis demonstrates, the legal system may use indiscriminate punishment
of transgressions as a means to deter unwanted behavior, whether arising
from the dominant norms of the law enforcement agency in question or
carried out at the officer's own initiative.

With model 7.3 in mind, we are ready to develop hypotheses.

H7.3.1. The relationship between the strength of societal norms and
the ability of officers to resist such prison norms, on the one hand, and,
on the other, the ability of judges to discern the guilt of individual officers is curvilinear. When the strength of such norms is extremely low or
extremely high, judges' ability to discern which officers commit inmate
abuse due to social pressure, and which instead engage in abuse due to
individual initiative, is relatively high.

The conditions of the separating equilibria in rows 2 and 10 of table
7.3.2 suggest that such equilibria are more likely if the s term is very
small or very large. If the norms are very strong and the judge observes
inmate abuse, he knows that the officers were forced to abuse prisoners
by the prevailing culture: the strong norms would have prevented the
high type deputies from inmate abuse. When the norms are very weak
and the judge observes inmate abuse, he can be certain that this is due to
individual initiative: such norms do not suffice to force low type deputies
to abuse prisoners.

H7.3.2. The prevalence of norms of inmate abuse across the penitentiary system increases the probability that inmate abuse will continue and
remain unpunished.

The conditions of a "dirty" equilibrium (rows 1, 4, 8 of table 7.3.2)
require that $\pi_0 > 1/2$. If the prevalent culture is abusive and individual officers cannot reasonably be held responsible for the inmate abuse, this creates additional incentives for the judge not to punish them. This also pro-
vokes undetectable unilateral abuse of prisoners.

H7.3.3. Under high indiscriminate punishment, prison officers will
refrain from inmate abuse if institutionalized norms of inmate abuse are
relatively rare or if the benefits of unilateral inmate abuse are high.

These are the conditions of a clean equilibrium, as listed on rows 6,

9, and 11 of table 7.3.2, and the separating equilibria in which the Judge can detect unilateral inmate abuse (rows 2 and 10). When the Judge reacts to any transgression as it were a conscious choice, not forced by any circumstances, and disregards the influence of the abusive norms, he makes the punishment imposed outweigh the ambient pressure, so that it reduces the incentives for low type deputies to follow the abusive norms and also the ability of high type deputies, who wish to engage in crime, to disguise themselves as low type deputies.

7.4. Conclusions

This chapter is the most complicated of those in which we have applied the algorithm for narratives and models. Each model here has been more challenging than any model devised in chapters 4 through 6, and within this chapter models advance in level of complexity. Because this increased complexity may appear as a major difference, we emphasize the epistemological commonalities between chapter 7, on the one hand, and, on the other, chapters 4 through 6. Throughout, particular theoretically driven questions have led us to construct distinct narratives from a single story, and then to design distinct models. Throughout, our choice of theoretical premises in each model has led us down different analytical paths. In each of the three sections, the models have yielded distinct equilibrium predictions, predictions that have successfully recovered their underlying, differing theoretical arguments.

Figure 7.4 summarizes the path that we have traveled in this chapter. As the figure indicates, we started with the *Times* story on police abuse of power in the L.A. County jail system. From that single story we extracted three narratives. In the first, to remind, the deputies see their unwritten code of conduct as essential, even though it is illegal. They end up being prosecuted and then punished for it. In this first narrative, the actors (A) are the deputies and the inmates and the behavior (B) examined is the level of abuse and the level of (anticipated) violence. Both actors have complete information (I). Recall as well that two equilibria (E) exist in this game. Whereas the first narrative focuses on the deterrence signal sent to inmates, the second highlights increasing transaction costs for Federal Agents.

Now that we have retraced the first model based on the first narrative, we simply observe that figure 7.4 guides reconstruction of the second and third models and the narratives on which they are based. More broadly, figure 7.4 illustrates our practical application in this chapter of the

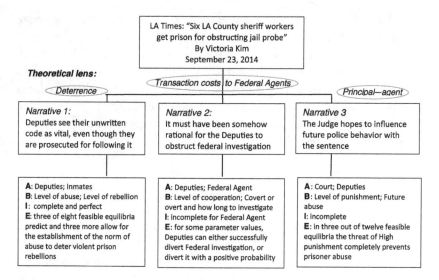

Key: **A**: Actors; **B**: Behavior; **I**: Information: **E**: Equilibrium

Figure 7.4. From one story to three narratives and then to three models

epistemological path introduced in chapter 2. It also summarizes how we use these epistemological principles while designing models and applying methodologies in chapters 4 through 7. Anyone so inclined can easily build for themselves analogous figures—and reconstructions of the paths from story to narratives and then to models—for each of chapters 4 through 6.

EIGHT

The Role of Modeling in How We Know What We Know

We design formal models in support of our theories because we believe that doing so improves our understanding of social phenomena and enriches the realm of accumulated theoretical knowledge. If we manage to deepen the disciplinary cognizance of how formalizations serve each of these objectives, we would engender even greater demand for cleaner, more deliberately designed models in the social sciences. Formal modeling commands respect as it is, and with proper epistemological placement it can deliver more fully on the high expectations that the community has for it.

A model is part of the process of theorizing, and yet is distinct from theory. It is situated at once after and before the two theoretical stages: it fuses the initial theory with limited and stylized empirical evidence, and "fits" a refined theoretical argument onto a slice of reality. Thus, a model supplies a consistency check for a theory. Whereas such a consistency check might be left to the theorist's process of intuitive thinking, or might even be impossible, as when, for example, in physics theorized phenomena are unobservable with current technology, for us modeling is both feasible and requisite. The reason why is that in the social sciences we do, on the one hand, witness social phenomena, and, on the other hand, those phenomena are of sufficient complexity to call intuitive comprehension into question. In the social sciences, at least, the strict mathematical logic of formal models is an assured improvement over unstructured intuition and belief.

Any superiority that formal models might have over intuition and speculation, however, only holds if mathematical accuracy and axiomatic transparency are nowhere violated. Without mathematical accuracy and transparency, we will merely have expressed the same intuitive understanding but in the vocabulary of mathematical concepts and notations. This is why the role we assign to formal modeling on the path to knowledge dictates that the method of modeling must meet specific requirements, and why in our comprehensive edifice the epistemology and methodology of formal modeling interdepend. The "how we know" and "how to" aspects of modeling must reinforce each other.

Entering into the edifice, so to speak, we designed models in response to theoretical queries using methods that meet the criteria of mathematical rigor and axiomatic transparency: basic utility maximization, decision-theoretic models, social choice theory, cooperative game theory, and non-cooperative game theory.

8.1. The Epistemological Place of Formal Models

The role of formal models in social research is to establish multiple linkages between and among theory, narrative, story, and evidence. The synthesis of the epistemology and methodology of modeling manifests when, via modeling, we answer theoretical queries as they are posed in the context of stories (observations). The displays in chapters 4 through 7 serve the purpose of highlighting these linkages and stressing the interrelated nature of the "how we know" and "how to" aspects of modeling.

The process of modeling is highly productive for our thinking about and making sense of social phenomena. It opens up multiple junctures for critical inquiries, review of premises, and generation and trying out of new ideas. Reconciling a theory with observation produces a plethora of ideas that are germane to that theory and enrich it. The chapters in which we have designed models in support of multiple theoretical arguments illustrate how fruitful this process is.

8.1.1. Models Test Theory

What does it look like, when, as we claim, models test theory? The chapter on the new award ceremony in an African village began by addressing the most obvious queries to arise from the one story that we used for inspiring the three basic utility maximization models in that chapter. The first model, which specified the utility function of the chief as a traditional officeholder, was spare in the extreme. In this, we followed our own advice, on starting with the simplest model possible as a baseline—we adhered to the theoretical principles embodied in the well-known Occam's razor. Spartan as it was, that first model (4.1) fulfilled the role that models play in testing theory in that it made explicit all of our assumptions about the Chief, and enabled us to check whether any of those assumptions were superfluous or inconsistent, violating the logicality of our theoretical statements.

Maintaining relative simplicity, our third model in that chapter (model 4.3) incorporated both the utility function of the Chief from model 4.1, as well as the elements of the utility functions of villagers from models 4.2.1 and 4.2.2. Model 4.3 validated its original theoretical argument while at the same time expanding that argument's predictions. Model 4.3 turned out to have substantive import in its own right: it identified the conditions under which villagers would—and would not—contribute to upholding a redefined traditional community in the contemporary era. More broadly, that model could now be tested, replicated, and used as a building block in other scholarship, thus contributing to knowledge accumulation.

Models in the chapter about pollock and salmon fisheries, while a different type than those in chapter 4, were also quite straightforward. They tested theory by, for example, compelling an explicit statement of assumptions about the commercial fishing business and examining the consistency and necessity of those assumptions. Furthermore, there we instantiated our argument about the role of formal models in the accumulation of knowledge. Recall that our first model in that chapter (model 5.1) entered as a building block for models 5.2.1 and 5.2.2. Model 5.2.2 then served as a building block within model 5.3. The findings of earlier models thus were incorporated as solved elements in the construction of subsequent models. Testable propositions in chapter 5 also differed from those in chapter 4: they emerged from the derived conditions on the parameter intervals within which specific predictions applied. To illustrate, model 5.3 predicted that the Fishery Management Council would issue a recommendation to change regulations for some parameter intervals, but it would not do so outside of those parameter intervals.

The chapter on heresthetic manipulation ahead of the Paris climate negotiations broke with our earlier pattern and addressed just one query about the strategy of British prime minister David Cameron for agenda setting as he was going into bargaining over the Paris Treaty. By then, we trusted the reader to agree that other queries could also be posed; and the query that we explored was but one of many possibilities. Thus, we dealt with the single query and the single narrative and delved into examining

188 Formal Modeling in Social Science

that narrative via two different methods. First, we modeled the situation using the tools of social choice theory, and then we modeled the same situation using cooperative game theory. While the axiomatic structure across the two models was similar, it could not be fully identical, since the difference in methods necessitated deciding on assumptions about different things: about preferences over alternatives in the social choice rendition of the narrative, versus about actors' absolute gains or losses vis-à-vis specific opponents on each dimension from a change in outcome in the cooperative game-theoretic rendition. The differences in assumptions implied differences in findings. Even so, in both renditions, the models tested theory: in both cases, we successfully aligned the datum of Cameron's speech with the theoretical framework of an agenda setter who engages in heresthetic manipulation to achieve his preferred policy.

In the chapter on the norm of inmate abuse in a Los Angeles prison, models were noncooperative game-theoretic. There we returned to the format of multiple queries and multiple narratives, because there were many theoretical possibilities to explore, and also because we wished to show the variation from simpler to more complex designs and from more general to more restrictive solution concepts (specifically, from Nash to perfect Bayesian Nash equilibria). We also showcased the process of moving back and forth on the choice of assumptions, for example, between rendering the same narrative as a model with complete versus with incomplete information.

Explicit in the iterative design of models in chapter 7 was the objective of obtaining the predictions that theory-based narratives suggested. Equally explicit was the work that we did in amending the assumptions when designing models. Admitting that amendments were necessary in order to reach the goal meant that we rejected the original theory due to its inability to logically justify the desired conjectures in its original form, prior to modifications. Thus we tested initial theories and accepted them only in their amended versions.

Testing theories via models is thus not a mystery at all. It is a common practice and merely requires acknowledgment and overt appreciation as a distinct epistemological step that we all take in our work. Of course, there is a snag: in the published version of our work, the original theory that a model has tested is left unreported, and only the amended version of a theory is presented. We cannot very well publish papers in the form of: "first I was thinking one way, but when building a model, I realized that it had to be this other way." Yet in a discipline as a community, we do see this process unfold in the progression from earlier to subsequent work and in the way that modelers critically assess their predecessors' arguments.

8.1.2. Models Build Theory

In recapitulating how models test theory, we have already edged into the discussion of theory-building. In fact, any change from the initial to the amended theoretical argument already constitutes theory-building.

The other instance of theory-building is when we take a prior argument as a block in constructing a more complex logical structure. The last model in the chapter on fisheries exemplifies how models perform that role. As emphasized, we depart from the resolved Prisoners' Dilemma as the standard model for enabling social cooperation in the face of the Tragedy of the Commons, as a fragment to include in all queries, all narratives, and all models in that chapter. Alongside the resolved PD, models fashioned early on in chapter 5 also enter as parts into subsequently designed models. Model 5.3 then combines the prior models as nested elements, using their predictions as an established platform in order to focus on the behavior of a new actor, the regulator (the Fishery Management Council).

Two well-known maxims contradict each other: "there is no honor among thieves" versus "there is honor among thieves." Based on a story about Los Angeles prison abuse, chapter 7 shows how either can be correct. The noncooperative game-theoretic models in chapter 7 are the most technically challenging in the book. Yet all of them are utterly intuitive. Because the reader can easily see and accept the intuition behind the argument even before it is successfully formalized, these models serve as perfect candidates to show the process of reconciliation between models and theory in action. We know what equilibria we need the model to produce; we just need to figure out how to achieve what we want. Hence we go through successive iterations when modeling the first narrative, and also by contrasting the implications of varying assumptions about information for the predictions in the games. The process of reconciliation between model and theory generates adjustments on both sides, and as the model is designed, the theory is changed. The resulting amended theory is better because we know it *can* work, as evidenced by the model's success.

8.2. Formal Models in Research Design

In addition to the contribution that models make within the epistemological chain, formalization can bring a number of specific tangible advantages to research projects, producing insights and clarity that scholars can find valuable. These are the benefits of formal models—extra strengths that they can bestow on research design. Incorporating formal models in research design enables theoretical replication, advances accumulation of knowledge, and not always but often enhances hypotheses.

8.2.1. Models Enhance Hypotheses

The formal models we design with readers in this book shed light on social behavior in ways that often align with extant relevant research. Even the simple models of chapter 4 yield testable implications resonant with the almost entirely nonformal literature on chiefly authority. Parenthetical cites in chapters 4 through 7 telegraph such connections with available scholarship that arise out of our work. That is to say, we have arrived at hypotheses that other scholars have tested. At the same time, our models also generate hypotheses that are new to the literature. This can be the case either because formal analysis manages to produce additional testable insights, or because some of our conjectures do not yield themselves easily to empirical investigation and are thus absent from the literature.

An important contribution of formal models that we have identified comes to the fore as we consider the discovery of conjectures that are not directly testable (cf., e.g., Clarke and Primo 2007, 2012). Without formal models, those types of conjectures are deemed nonfalsifiable, and fall by the wayside, dismissed by the scientific method. When derived as one or several among many from a single formal logical structure, such once non-falsifiable conjectures, however, can be affirmed. Insofar as some of the linked conjectures yield themselves to direct evidentiary assessment, their companion propositions can be viewed as also having been subjected to testing by implication.

A special forte of formal models is producing counterintuitive insights. Many introductions to game theory begin with the Prisoners' Dilemma specifically because to a novice the findings appear counterintuitive. Of course, what is counterintuitive for one person is but a theoretical claim for another: under the assumption of methodological individualism, the common good theory does not work, and instead the PD's prediction of breakdown of beneficial cooperation applies. Even so, this ability to discover the nonobvious is invaluable in formal models. Fiorina (1975) discusses how a formal model generated a result contrary to suggestions about empirical patterns prevalent in the available literature. As Fiorina (1975, 145) puts it, "Working with the model disabused me of some notions that both casual argument and earlier studies seemed to support." The expectation from such a model, as he also observes, subsequently found empirical support. Hypotheses that such models would suggest would differ sharply from common sense and the commonplace.

8.2.2. Models Enable Theoretical Replication

We contend and demonstrate in this book that formal models enable theoretical replication. A model's accuracy is verified and its full set of assumptions is scrutinized each time a reader examines a piece of scholarship that contains a model. Consider what an equivalent to that might look like in empirical research. This would be like checking every single datum and its source, as well as the appropriateness of the estimation method and each of its assumptions, and then rerunning the estimations whenever a piece including statistical analysis is read. With statistical analysis, this scenario simply is not feasible: most readers must take the findings on faith. Indeed, a repeat investment of effort of comparable magnitude to that for the original research would be required if a reader wanted to fully replicate statistical findings. For the obvious reasons of high effort and low return, as replication in and of itself rarely produces publication, this is rarely done.[1]

The role of formal modeling in theoretical replication is unique and heretofore underappreciated. It is vital, as it enforces scholarly accountability. Since the 1980s, political scientists and social scientists more broadly have stressed the role, need, and importance of replication in empirical, but not in theoretical, work (e.g., Bollen et al. 2015; Dewald, Thursby, and Anderson 1986; Gilbert et al. 2016; King 1995; Lupia 2008; Lupia and Alter 2014). The contributors to this tradition focus for the most part on enforcing easy replicability of published results as well as creating incentives for scholars to engage in replication of others' findings insofar as statistical analysis is concerned, that is, in observational and experimental methods.

The epistemological role of formal models in serving as platforms for theoretical replication is akin to how statistical models enable evidentiary replication. We want to have confidence rather than faith in the author's theoretical argument, in particular when that argument is complex and when the conditionalities on which claims are based are many. In the spirit of enabling replication, then, we advocate as a standard that theoretical claims are logically formalized whenever possible.

Beyond their power to supply theoretical replicability, formal models have synergy with empirical replication. The word "synergy" can be over-used, yet here it is apt. Because of the power of models to logically link together several conjectures, they broaden the range of means for replicating statistical findings. Whereas scholars frequently seek to replicate a finding on a new or expanded source of data, obtaining identically measured variables may not be feasible. Insofar as their hypotheses are linked via a formal model, a scholar need not limit herself to identical operationalizations. Her opportunities for replicating results and testing theories thus broaden substantially to include a wide variety of contexts. Instead of a single operationalization, multiple operationalizations *linked* via a model will be obtainable for a larger set of subjects, or countries, or historical periods.

8.2.3. Models Advance Knowledge Accumulation

When scholars use an extant formal model, whether long-standing or newly designed, in constructing new models to answer new queries, they can be confident that the elements they thus incorporate are logical and logically replicable. They also know that, for these reasons, as long as they continue making consistent assumptions, the building blocks made of extant models should—and will—work in a known way in explaining and predicting actors' behavior. This process exemplifies the use of accumulated knowledge.

The value of falling back on the shared accumulated body of knowledge (for present purposes, formal models in published scholarship) goes beyond expediting the production of new models. It generates axiomatic consistency in linked research, and, relatedly, axiomatic distinctiveness, when the purpose of new models is to answer apparent empirical puzzles. Furthermore, utilizing shared knowledge carries great communication value. Inclusion of substructures that are shared knowledge in the discipline facilitates the reader's comprehension of the overall logic of a new design, and serves to highlight which aspects of it are novel to the field.

In using prior results as integral building blocks for new analyses, scholars thus both rely on accumulated knowledge and add to the stock of knowledge. Any properly constructed formal model can be reused as a building block—as long as its premises remain unchanged. In that sense, any such model contributes to existing knowledge. This is not to say that models' usefulness does not vary widely. Some will become famous, while others will remain esoteric. But even an esoteric model can be pulled out of storage at any time and included as a part of a new design with full assurance of its logical consistency. Thus, formal models enable accumulation of knowledge.

8.3. Our Epistemological Claims

Throughout this book, we advance a number of epistemological claims. We want once again to draw the reader's attention to them and their importance. These are what we consider the takeaways about what formal models are in epistemological terms, and why they are not a supplementary, clarifying option in science but rather an integral part of it.

To spotlight, our epistemological claims are:

- The primacy of theory for raising a query: $T \rightarrow N$
- The primacy of a theory-driven query for constructing a narrative: $T \rightarrow N$
- The primacy of a narrative in establishing assumptions: $N \rightarrow M$
- The model as well as the narrative each fuses theory with observation: $E_i \rightarrow N \leftarrow T$ and $E_i \rightarrow M \leftarrow T$
- The logicality of a model tests the validity of theoretical conjectures: $M \rightarrow T'$
- Model and theory get reconciled through amendment of assumptions: $T \rightarrow M \rightarrow T'$
- The role of evidence in theory building: $E_i \rightarrow M \rightarrow T'$
- Evidence enters before and after modeling, in a dance between reality and abstraction: $E_i \rightarrow M \rightarrow E$

These bullet points harken back to the epistemological schema in chapter 2, which outlined our approach to the epistemology of social science research as it incorporates formal modeling. Along the path throughout the book, we have developed these ideas and illustrated them by designing models with multiple linkages to theory and evidence. Taking up the first bullet point, we observe that the researcher has some theoretical concern in mind when she poses a query about a phenomenon ($T \rightarrow N$). The theory-driven query she frames shapes the implicit or explicit narrative she constructs to make sense of the phenomenon under study, again ($T \rightarrow N$). The narrative in turn undergirds the assumptions driving the model she designs ($N \rightarrow M$). The logicality of her model tests the validity of the theoretical propositions with which she started. Establishing that logicality may require amending her original theory ($M \rightarrow T'$). Through this process, formal modeling makes possible the definitive achievement: we can reconcile model and theory by amending our assumptions ($T \rightarrow M \rightarrow T'$). In identifying what needs to be amended, the model tests the old theory, T, and in adding assumptions to arrive at the new theory, T', the model, or rather the process of modeling, builds theory. As emphasized, real-world phenomena, even if only implicitly, guide us in our design of models and guide, too, the amendment of theory (E_i → M → T'). Scholars follow what amounts to an ongoing dance of reality and abstraction: evidence enters before and after modeling (E_i → M → E), as we move from an element of reality, to designing the model, to explaining an entire class of social phenomena.

This review of our epistemological claims reiterates the understanding that scholars always have an implicit narrative in their heads whenever they approach a research question grounded in actual social phenomena. The query—with its accompanying narrative—might regard interactions in marriage, the conditions for economic growth, the workings of federalism, or the impact of demographic change on labor markets or housing or health policies. Building a model with explicit assumptions and identifying its mathematical representation clarifies and enforces consistency among all parts of the analyst's approach to answering her query. Once again, modeling requires the analyst to make explicit the ingredients of her reasoning and logically trace through all the steps between assumptions and conclusions.

We have seen the many advantages of designing models that are original and not pulled off the shelf. Attempting to "fit" phenomena to existing models in fact contradicts our epistemological understanding: analysts may well compromise their queries in trying to squeeze them within the confines of extant models. Instead, we argue that analysts would be well advised to ponder not which extant model might approximate their puzzle, but rather whether their ideal new design can benefit from incorporating any of the known models as elements.

8.4. The Meta-Method of Formal Modeling

Having revisited our epistemological claims, we now set out the book's main methodological claims. We have two sets of methodological observations. The first regards choosing a method, and the second reflects the necessarily dynamic and active nature of modeling. These do not correspond to the conventional notion of methodology as a toolkit. The toolkits instead are the methods we surveyed in figure 3.1, each with plenty of rules and complexity, requiring training and time to implement. Each toolkit is under continuing development as practitioners strive for improvements. Toolkits are not what we highlight. Rather, we emphasize the meta-characteristics of the formal modeling methodology, which are inextricably joined to the fulfilment of its epistemological function.

Choosing a method

- There are multiple methods in economic modeling, distinguished by their axiomatic structures and solution concepts.
- The choice of method means accepting the set of basic assumptions associated with that method, and thus has implications for the predictions of a model. The researcher cannot, within that method, attempt to change basic assumptions.
- The choice of method depends on which set of basic assumptions the researcher sees as the most apt.

Methodology of iterated model design

- Reconciliation between theory and model requires amending the axiomatic structure of both: a model is adjusted until it recovers, with all of its moving parts working together, the outcome that a theory should produce.
- Before deriving predictions, the analyst must establish *Existence* of the solution.
- When *Existence* obtains, the analyst must check *Logicality*.
- Solutions exist for specific parameter intervals, and predictions vary across parameter values.
- Untestable propositions can be logically linked, via a model, to testable conjectures.
- Propositions requiring diverse supporting operationalizations can be logically linked via a model.

A particular model or set of models, whether extant or newly designed, constitutes a tool for understanding and discovery. The knowledge of how to design models, assess models, and reconcile theory and model constitutes, in turn, what might be termed a meta-method, straddling the boundary between epistemology and methodology.

8.5. Broad Implications and the Agenda Looking Forward

Our epistemological argument as well as its methodological implementation have broad implications for social research. We spell out what scholars practice but not discuss. Scholars do not and need not merely theorize and then discard in their progression to knowledge. Instead, they theorize iteratively, and discard only as the last resort. Theories evolve, getting fused with moderate doses of evidence, even before and in preparation for the full-scale confrontation with the data as in statistical hypothesis testing. Before we test our thinking, we model—and as we model, we refine our thoughts. We "reality check" our arguments. And sometimes, depending on the argument's subject and our observational capacity, such reality checking is the limit of our ability to critically evaluate a theory. Even if that is the case, a model in its own right provides a means for critical evaluation.

Another broad observation of paramount importance is the centrality of a narrative, which we affirm for social scientific epistemology. The pervasiveness of explicit and implicit narratives evidences that we do have, and must have, a synthesis of observation and theory within the process of theorizing. While narratives do not conclusively and replicably contribute to theorizing on their own, they prepare the raw materials for modeling and spotlight the nature of a model as a synthesis of theory and observation.

Third, many of our friends and colleagues have raised an intriguing question: Can we dispense with the "art" in the "art of modeling" and entrust modeling and theory revision to a computer, as we did long ago with statistical methods? We say that the iterated algorithm for minimally amending theory in the construction of a workable model can feasibly be mechanized. A computer can process permutations of assumptions. The art would likely be still required, however, as processing the characteristics of complex social phenomena into discrete analytic bits as well as selecting which assumptions can and cannot be meaningfully amended would always require active intelligence.

Our claims and their implications enlarge the agenda for the use of formal modeling in social science. Throughout this book, we plumb the depths of deductive reasoning. We also highlight the role of induction: we repeatedly stress that all students of human behavior start at least implicitly with a narrative about the social phenomenon they have in mind. We mention abduction, too, representing it as an extreme case of induction, which occurs when explaining the observed phenomenon is the singular goal of modeling. The difference between the inductive and abductive method captures the difference between our approach and the method of analytical narratives. Indeed, this difference is one of degree, and in chapter 6 our two models explain the occurrence of a unique, likely unrepeatable, event.

As we move through the process of designing formal models, we expose the occasional clash between positive predictions and normative assumptions or expectations. The relationship between formal theory and normative theory remains in the background throughout our journey, as indeed we refrain from commenting on how unfortunate and just plain wrong it is that norms of abuse arise rationally in prison communities. Instead, we make positive predictions via a model that such norms should arise under specified conditions. Many prominent political theorists, however, have brought this dissonance between positive and normative arguments to the forefront of the debate (e.g., Ferejohn 2007). The most notable example of such disagreement is the revision of the long-held premise of the existence, even primacy, of "the will of the people." Irreverently, but mathematically irrefutably, in a string of fundamental results in social choice theory, cooperative game theory, and noncooperative game theory, formal social scientists have argued that, as conventionally understood, the will of the people amounts to a will o' the wisp (e.g., Arrow 1951; McKelvey 1976, 1979; McKelvey and Schofield 1986; Olson 1965; Riker 1982). To refute their claims, one would have to refute their assumptions, since the logical bridges in their work are transparent and replicable. Of course, since most of their assumptions are the assumptions of modern economics, there have not been many attempts to discredit these results.

Failing to replicate a theoretical argument via a formal model, such as in the example just named, need not spell the end for the argument, nor does it discredit the formal method. Instead, it might suggest the omission from our theoretical constructs of some additional social mechanisms serving to buttress the desired outcome in the face of the logically revealed challenges to achieving it. A case in point is Mancur Olson's (1965) exposé of the impossibility of social cooperation—until and unless supporting institutions and structures are put forth by the communities to enable it.

A focus on theoretical replication in the social sciences is an innovation that can affect how and when we include, or even require, logical models. To date, replication in the political science and social science scholarly community has been overwhelmingly understood in empirical (observational and experimental) terms.[2] Political scientists have promoted empirical replication in a recent push for data access and research transparency (DA-RT) (e.g., Carsey 2014; Elman and Kapiszewski 2014; Hartzell 2015; Htun 2016; Isaac 2015; Lupia 2008; Lupia and Alter 2014; Lupia and Elman 2014; Schwartz-Shea and Yanow 2016). Scholars in other disciplines have made important contributions to furthering empirical replication in the social sciences (e.g., Anderson and Dewald 1994; Bollen et al. 2015; Camerer et al. 2016; Dewald, Thursby, and Anderson 1986; Freese 2007; Fuess 1996; Hamermesh 2007; Hubbard and Vetter 1996; Kane 1984; Nosek et al. 2017).

The discussion of theoretical replication and its role in the accumulation of knowledge advances this broad agenda of promoting replication.

Moreover, theoretical replication and the formal method specifically can supply a synergy with empirical replication, as when the formal method permits logically linking several propositions, a subset of which are testable hypotheses. The understanding and practice of replication in social science must expand if we are to seek scientific advances where direct empirical testing is not feasible, spur the debate on the integration of theoretical and empirical research, and multiply venues for the accumulation of knowledge in political and social science. Theoretical replicability and accumulation of theoretical knowledge are inseparable.

The agenda we identify is not exhaustive. That would be impossible, given the nature of human inquiry in science. We deliberately set out a selective agenda for the use of modeling to enhance the study of human behavior. We choose what we see as the most important aspects, given the current state of the art of modeling social processes.

Human knowledge is limited, and any prediction of the future we might venture applies in full only in the imaginary universe that is spanned by our own starting premises. Precisely because of the nature of human scientific knowledge, imperfect yet cumulative, this book is not about answers, but about ways of posing queries and searching for answers. Any scholar in any field, be it in the arts and humanities, the natural and life sciences, or the social sciences, seeks discovery and innovation. What we have shown for the study of social behavior is that formal modeling serves as an engine of discovery and innovation, driving forward the production and accumulation of new knowledge.

Notes

Chapter 1

1. We thus revisit Fiorina's (1975) celebrated piece for a discipline in which mathematical social science has become relatively widespread and is no longer a monopoly of "technical people."

2. Observe that even a single word, shared in common, can reveal distinctions between statistics and formal modeling. Whereas in statistics, a parameter refers to a variable characterizing a population with a value that the researcher seeks to recover as she analyzes evidence from samples, in modeling (as in mathematics more broadly) parameter means an argument of a correspondence or a function.

3. "Logicality" is used here as it is used in, for example, physics and the philosophy of science to denote not logic, precisely, but rather the ability to meet conditions of logic when progressing from premises to conclusions (e.g., Coecke and Smets 2004; Stuewer 1998).

4. See, e.g., Abbott 1992, 2007; Hymes 2003; Hyvärinen 2016; Labov 2003, 2006; Maines 1993; Nash 2005; Reck 1983; Richardson 1990; Skultans 2008; Strawson 2015; Toolan 2012; Yamane 2000.

5. Chapter 2 discusses the difference between the concept of narrative as a stage in research design and the distinct methodology of analytic narratives represented in the work of such scholars as the renowned team led by Robert Bates (1998, 2000a, 2000b) and those who have adopted an analogous approach.

6. For a comprehensive review of the use of narratives across disciplines, see Patterson and Monroe 1998.

7. Notably, Kramer (1986, 12) agrees: "Prediction and control are often useful byproducts (and the ability to generate falsifiable predictions which can be tested is of course a vital reality test), but the central object is understanding: being able to organize or 'explain' a range of empirical phenomena by finding the underlying principles or laws which account for them."

8. For different and divergent positions regarding modeling and hypothesis testing, see, e.g., Clarke and Primo 2007, 2012; Johnson 2010; and Morton 1999.

9. On transparency in empirical research, see, e.g., Isaac 2015; Lupia and Elman 2014; and Schwartz-Shea and Yanow 2016.

10. This book may also be of benefit to political and social science graduate students as they decide whether it is worthwhile to pursue (further) formal theory training given the place of modeling in social research.

Chapter 2

1. To see the link from T to M: observe that T → N → M, as discussed below and as represented with the horizontal arrow in figure 2.1.

2. This holds if the method we use is noncooperative game theory. If we use another type of model, solution concepts other than the Nash equilibrium apply. To clarify, a solution concept denotes a rule for what would be considered as satisfying the criteria for being a prediction from the model. We return to solution concepts in chapter 3. In an (Nash) equilibrium, no actor has an incentive to change strategy unilaterally, while others' strategies stay the same. For details, see chapter 3.

3. See Clemens 2017 for an overview of empirical replication (vs. robustness tests) as understood and practiced in six disciplines. In political science, the literature on empirical replication is ample and growing (e.g., Bueno de Mesquita et al. 2003; Dafoe 2014; Green, Palmquist, and Schickler 1998; Ishiyama 2014; King 1995; Lieberman 2010; Neely 2007; Ward 2004), as it is in other social sciences (e.g., Bollen et al. 2015; Gilbert et al. 2016; Gómez et al. 2014; King 2011; Hunt 1975; Martin and Sell 1979; McNeeley and Warner 2015). A range of scholars in multiple social science disciplines understand theoretical replication as entailing moves in empirical analysis (Aligica and Evans 2009; Bahr et al. 1983; Christensen 2006; Hyman and Wright 1967; La Sorte 1972; Parkhe 1993; Perry 1998; Whelan and Teigland 2013; 2011, 2017).

4. If the scholarly wisdom suggests that something is true, but formal theory reveals no way to support the wisdom, then that in itself is important. Indeed, the Prisoners' Dilemma itself is evidence of the nonlogicality of the common good theory!

5. Pareto efficiency is defined as a situation where it is not possible to improve the lot of any individual without making at least one other person worse off.

6. A mixed strategy Nash equilibrium is a Nash equilibrium where each player is permitted to use linear combinations of all their strategies.

7. This is one manifestation of Rudolf Carnap's "logical probabilities." As he states: "Logical probability is a logical relation somewhat similar to logical implication; indeed, I think probability may be regarded as a partial implication" (Carnap 1987, 168).

8. Facts enter into a social scientist's effort to generalize in both theorizing and modeling. But whereas theories generalize from observation, models reconstruct the observed fact, while treating it as an element of some presumed more general class.

9. We exclude the possibility of cheating, where there are contrary bits of evidence that are suppressed; we discuss the ideal case.

10. Chapter 8 returns to the distinction between inductive and abductive reasoning, one made within philosophy of science, logic, and artificial intelligence (e.g., Flach and Kakas 2000; Flach and Hadjiantonis 2013; Hempel 1965; Peirce 1958). Note that Bates and colleagues (1998, 3, 241; cf. 2000) label their analytic narratives approach as a methodology, while we view it as epistemology.

11. See Putnam (1991) on applications in physics as confirming or disconfirming theory. Social scientists have similar experiences in the fields of, for example, public policy, program design, and the growing emphasis on programs' assessment.

Chapter 3

1. An individual has *transitive* preferences if she prefers option A to B, B to C, and then it also follows that she prefers A to C.

2. Tsebelis (1989) clearly delineates the difference in choosing to model a social phenomenon within decision-theoretic versus game-theoretic frameworks.

3. To grasp the invulnerability of the outcomes in the core, consider that, the core, if it exists, is an intersection of the Pareto sets of all decisive coalitions: any move away from any policy in the Pareto set will be disliked by at least one member of that decisive coalition. Thus, any proposal to amend policy away from an outcome in the core would lose a vote.

4. Likewise, the same phenomenon of Supreme Court decision making is analyzed within a cooperative game-theoretic framework in Epstein and Mershon (1996) and a noncooperative game-theoretic framework in Epstein and Shvetsova (2002)—to different ends.

5. Without rigor and clarity in the use of the method, a model will not credibly identify logical limitations and inconsistencies in the original theory. Nor will it produce a reliable "fix" amending the original theory. Unless we are certain that the mathematical apparatus, however minimal, is accurately applied, what we have is not in fact a model in the epistemological sense we discuss. Instead, such non-models can be seen as simply extensions of informal theoretical discussion invoking some mathematical vocabulary.

Chapter 4

1. Note what this progression implies: we do not start from the available scholarly literature. We work from the world as depicted in the story, and we take up the interactions among people as described in the story. We go next to narrative, and from that move on to model. Only then do we connect with extant literature.

2. Source: Phandle 2014. South African spelling is retained. http://www.dispatchlive.co.za/news/2014/06/06/mbhashe-chief-to-reward-trailblazers/

3. A minimalist definition identifies chiefs as unelected local leaders who belong to the communities they rule and who, once elevated, rule for life. Indeed, in Xhosa,

nkosi is a title of a chief, lord, or king. Note that this definition and the discussion here entail assumptions about a chief as an officeholder.

4. Here, as in chapters 5–7, the assumptions at the base of a given model come from the world as depicted in the story. If the contents of the story do not "signal" all of the assumptions needed, we either fall back on external sources or make assumptions ad hoc. As it happens, case studies in South Africa and sub-Saharan Africa more broadly indicate that the motives just described are common among senior traditional leaders (e.g., Baldwin 2015; Oomen 2005; Williams 2010; for a less sanguine view, see Ntsebeza 2005).

5. As intended, narrative 4.2 reiterates a portion of the story at the outset of this chapter. We extract elements of the original story for one narrative—of the multiple narratives possible—and subject those elements to sustained, separate analysis.

6. Equivalently, we could also say that they believe the returns from market agriculture to be too low, or both the returns to be low and the probability of failure to be high, but we choose to capture their unwillingness to diversify their assets as caused by their risk aversion and perception of risk as very high.

Chapter 5

1. Here, as before, we start with a story, craft narratives, and make models, so following the schema as presented in figure 2.1.

2. By "baseline," we emphasize that nothing new is created in this portion of our analysis, as the reader will soon see.

3. "Fisher" is an accepted gender-neutral term in the industry (Branch and Kleiber 2015).

4. The regulatory tool of an individual transferable quota allocates to individual fishers the right to fish a certain amount of the target species each year. That is, these quotas work toward ensuring the sustainability of fisheries by endowing fishers with a property right over fish even before they catch it.

5. Solving for F: $[10 > 15–F \rightarrow F > 5]$ and $[5 > 7–F \rightarrow F > 2]$. The larger of the two constraints is 5, and that is the one that needs to be met or exceeded in order to shift equilibrium behavior.

6. The assumption is consistent with the commercial fishers' claim that only a small portion of their Chinook bycatch is on its way to the Kuskokwim and Yukon River deltas.

7. This general preference to preserve species other than the staple is not necessary to make our argument. In fact, this feature creates an unfavorable bias that the model needs to overcome if it is to recover as its prediction the behavior that we expect from the narrative, namely, that commercial fishers choose aggressive technology. As with statistical modeling, in formal modeling, if we are aware of introducing any bias in our work, we strive to bias our model *against* what we expect.

8. To remind, α denotes excessive bycatch, and β the probability of Chinook survival, conditional on high bycatch.

9. The owner of a fleet might respond to the incentive she faces given monitoring and enforcement under a regulatory regime. Yet suppose that the owner hires

multiple fishers to operate the fleet. A potential for agency loss now emerges: each of the fishers employed maximizes his or her own catch, and jointly they overfish—exceed the allotted quantity of—either the main staple or, as here, the bycatch. Since jointly they do not bear the same direct cost of overfishing as the owner of the fleet does, they do not have the same incentives to be cautious. This is why the regulator would want to position monitors on individual ships.

10. For example, others could design a model in which the more stringent regulation creates a secondary collective action problem, when all commercial fisheries use Aggressive technology, the hard cap is reached, and all fishing must cease. They would collectively all be better off with moving to use Conservative technology, but their individual rationale pushes them to act otherwise.

11. Note the simplifying assumption that the use of Conservative technology would avoid hard caps on Chinook bycatch, and so would allow fishing businesses to complete their allotted quotas of pollock.

12. Specifically, the NPFMC, one of eight such U.S. councils, is made up of members nominated by the governors of Alaska and Washington and appointed by the secretary of the U.S. Department of Commerce, as well as fisheries officials from Alaska, Washington, and Oregon, and the Alaska regional director of the National Marine Fisheries Service. Nonvoting members come from the Pacific States Marine Fisheries Commission, the U.S. Fish and Wildlife Service, the U.S. Coast Guard, and the U.S. State Department (NPFMC 2017).

13. A subgame is the set of all nodes, branches, and payoffs that follow after a singleton information set and without breaking any further information sets, for example, on the left of figure 5.2.2. Actors (and analysts) who roll back, or, equivalently, engage in backward induction, anticipate final outcomes and behave accordingly from the outset.

Chapter 6

1. Available at https://www.youtube.com/watch?v=f1NZEqgd7zw. The video also had few likes (eighteen) and only one dislike.

2. Available at https://www.youtube.com/watch?v=WF0twf3u5wA and https://www.youtube.com/watch?v=s2CeDSR6rz8, respectively.

3. Note that here and below, we abbreviate the labels (between parentheses) for packaged alternatives and preferences over them. To illustrate, we abbreviate the last packaged alternative as "(No Paris, but free trade)" when it actually contains, as the full listing to the left displays: no Paris, no EU accord, but free trade with subsidies.

4. Observe that a preference ordering as captured just now is clearer than that expressed in a vertical listing, for the vertical appearance can easily suggest a strict ranking among all preferences.

5. Indeed, in a critique of the process that led to the Paris agreement, Lumumba Di-Aping, a lead negotiator for Non-Annex I countries, referred to the negotiations in the following way: "You cannot ask Africa to sign a suicide pact, an incineration pact, in order to maintain the economic dominance of a few countries" (as quoted in Chin-Yee 2016, 359).

6. An Edgeworth box is an analytical tool to represent two actors trading two commodities. It depicts actors' preferences over a space defined by two axes, one for each commodity. Here we use a modified version of an Edgeworth box, where one of the goods is strictly distributive (the horizontal, economic axis), while the other is jointly increasing (the vertical, environmental axis). The slant in the label points at the direction of this joint increase.

Chapter 7

1. We use this game as a building block in a subsequent model, in which the Guards choose to institutionalize their behavior.

2. The reader can contemplate a number of alternative games reflecting the narrative. One possibility is the collective action problem involved in the implementation of the unwritten code: while the Sheriff's Department as a whole might benefit from abusive norms, an excessive use of force need not be individually rational; this might lead to an equilibrium in which some officers abide by the norm, and others do not. Another possibility is to model the norm of inmate abuse as a repeated game with incomplete information about the costs of repression.

3. Bayes's rule requires that actors update their beliefs whenever relevant information becomes available in accordance with conditional (posterior) probability formula.

4. In a separating equilibrium, actors of different types play observably different strategies.

5. In pooling equilibria, actors of different types choose identical action, and thus are observationally indistinguishable.

Chapter 8

1. But see Boix and Stokes 2003 for a highly cited example of replication with the addition of several robustness tests. Additional such examples do not undermine the point that replications alone are rarely published.

2. The wisdom is well captured in the 2015 report on robust and reliable research prepared by the Advisory Committee to the Directorate for Social, Behavioral, and Economic Sciences (SBE) of the National Science Foundation (Bollen et al. 2015). Building on that basis, the NSF Dear Colleague Letter on Robust and Reliable Research in SBE (NSF 16–137) invited principal investigators to submit research proposals on advances in empirical replication.

References

Abbott, Andrew. 1992. "From Causes to Events: Notes on Narrative Positivism." *Sociological Methods and Research* 20, no. 4: 428–55.

Abbott, Andrew. 2007. "Against Narrative: A Preface to Lyrical Sociology." *Sociological Theory* 25, no. 1: 67–99.

Acemoglu, Daron, Tristan Reed, and James A. Robinson. 2014. "Chiefs: Economic Development and Elite Control of Civil Society in Sierra Leone." *Journal of Political Economy* 122, no. 2 (April): 319–68.

Adams, James F., Samuel Merrill III, and Bernard Grofman. 2005. *A Unified Theory of Party Competition: A Cross-National Analysis Integrating Spatial and Behavioral Factors.* New York: Cambridge University Press.

Agriculture and Rural Development Department. 2004. "Saving Fish and Fishers: Toward Sustainable and Equitable Governance of the Global Fishing Sector." Publication no. 29090-GLB. Washington, DC: World Bank, Agriculture and Rural Development Department.

Akcinaroglu, Seden, and Elizabeth Radziszewski. 2005. "Expectations, Rivalries, and Civil War Duration." *International Interactions* 31, no. 4: 349–74.

Akcinaroglu, Seden, and Elizabeth Radziszewski. 2013. "Private Military Companies, Opportunities, and Termination of Civil Wars in Africa." *Journal of Conflict Resolution* 57, no. 5: 795–821.

Aldrich, John H. 1995. *Why Parties? The Origin and Transformation of Political Parties in America.* Chicago: University of Chicago Press.

Aldrich, John H. 2011. *Why Parties? A Second Look.* Chicago: University of Chicago Press.

Aldrich, John H., and James E. Alt. 2007. *Positive Changes in Political Science: The*

Legacy of Richard D. McKelvey's Most Influential Writings. Ann Arbor: University of Michigan Press.

Aligica, Paul Dragos, and Anthony J. Evans. 2009. "Thought Experiments, Counterfactuals and Comparative Analysis." *Review of Austrian Economics* 22, no. 3: 225–39.

Allison, Edward H., Blake D. Ratner, Björn Åsgård, Rolf Willmann, Robert Pomeroy, and John Kurien. 2012. "Rights-Based Fisheries Governance: From Fishing Rights to Human Rights." *Fish and Fisheries* 13, no. 1: 14–29.

Alt, James E., Randall L. Calvert, and Brian D. Humes. 1988. "Reputation and Hegemonic Stability: A Game-Theoretic Analysis." *American Political Science Review* 82, no. 2: 445–66.

Alt, James E., and John T. Woolley. 1982. "Reaction Functions, Optimization, and Politics: Modelling the Political Economy of Macroeconomic Policy." *American Journal of Political Science* 26, no. 4: 709–40.

American Civil Liberties Union. 2011. *Banking on Bondage: Private Prisons and Mass Incarceration.* New York: ACLU. https://www.aclu.org/banking-bondage-private-prisons-and-mass-incarceration

Amnesty International. 2014. *Entombed: Isolation in the US Federal Prison System.* Publication no. AMR 51/040/2014. https://www.amnestyusa.org/reports/entombed-isolation-in-the-us-federal-prison-system/

Anderson, Richard, and William G. Dewald. 1994. "Replication and Scientific Standards in Applied Economics a Decade after the Journal of Money, Credit and Banking Project." *Federal Reserve Bank of St. Louis Review* (November): 79–83.

Ansolabehere, Stephen, and James M. Snyder. 2000. "Valence Politics and Equilibrium in Spatial Election Models." *Public Choice* 103, no. 3: 327–36.

Arendt, Hannah. 1970. *Men in Dark Times.* Orlando: Harcourt, Brace & Company.

Arnason, Ragnar. 2012. "Property Rights in Fisheries: How Much Can Individual Transferable Quotas Accomplish?" *Review of Environmental Economics and Policy.* 6 no. 2 (August 16): 217–36.

Arrow, Kenneth J. 1951. *Social Choice and Individual Values.* Hoboken, NJ: John Wiley & Sons.

Association of Village Council Presidents. 2016. "AVCP Unit Representative and Chairman Election Results." October 11. http://www.avcp.org/avcp-unit-representative-and-chairman-election-results/

Association of Village Council Presidents. 2017. "Association of Village Council Presidents." http://www.avcp.org/

Austen-Smith, David, and Jeffrey Banks. 1990. "Stable Governments and the Allocation of Policy Portfolios." *American Political Science Review* 84, no. 3: 891–906.

Austen-Smith, David, and Jeffrey S. Banks. 1999. *Positive Political Theory.* Ann Arbor: University of Michigan Press.

Axelrod, Robert, ed. 2015. *Structure of Decision: The Cognitive Maps of Political Elites.* Princeton: Princeton University Press.

Bahr, Howard M., Theodore Caplow, and Bruce A. Chadwick. 1983. "Middletown III: Problems of Replication, Longitudinal Measurement, and Triangulation." *Annual Review of Sociology* 9, no. 1: 243–64.

Balalaeva, Dina. 2012. "Innovations as Public Goods Provision with Negative Externalities: Role of Parliamentarism." *Higher School of Economics Research Paper.* No. WP BRP 06/PS/2012. Available at SSRN: http://dx.doi.org/10.2139/ssrn.2192027

Balalaeva, Dina. 2015. "Political Competition, Agenda Power, and Incentives to Innovate: An Empirical Examination of Vested-Interest Theory." *Review of Policy Research* 32, no. 4: 413–42.

Baldwin, Kate. 2013. "Why Vote with the Chief? Political Connections and Public Goods Provision in Zambia." *American Journal of Political Science* 57, no. 4: 794–809.

Baldwin, Kate. 2014. "When Politicians Cede Control of Resources: Land, Chiefs, and Coalition-Building in Africa." *Comparative Politics* 46, no. 3: 253–71.

Baldwin, Kate. 2015. *The Paradox of Traditional Chiefs in Democratic Africa.* New York: Cambridge University Press.

Banks, Jeffrey S., and D. Roderick Kiewiet. 1989. "Explaining Patterns of Candidate Competition in Congressional Elections." *American Journal of Political Science* 33, no. 4: 997–1015.

Baron, David P., and John A. Ferejohn. 1989. "Bargaining in Legislatures." *American Political Science Review* 83, no. 4: 1181–1206.

Barrett, Scott. 1999. "A Theory of Full International Cooperation." *Journal of Theoretical Politics* 11 (4): 519–41.

Bates, Robert H., Avner Greif, Margaret Levi, and Jean-Laurent Rosenthal. 1998. *Analytic Narratives.* Princeton: Princeton University Press.

Bates, Robert H., Avner Greif, Margaret Levi, Jean-Laurent Rosenthal, and Barry Weingast. 2000a. "Analytic Narratives Revisited." *Social Science History* 24, no. 4: 685–96.

Bates, Robert H., Avner Greif, Margaret Levi, Jean-Laurent Rosenthal, and Barry R. Weingast. 2000b. "The Analytic Narrative Project." *American Political Science Review* 94, no. 3: 696–702.

Bayley, David H. 2002. "Law Enforcement and the Rule of Law: Is There a Trade-off?" *Criminology and Public Policy* 2, no. 1 (November): 133–54.

Beall, Jo. 2005. "Exit, Voice and Tradition: Loyalty to Chieftainship and Democracy in Metropolitan Durban, South Africa." Crisis States Research Centre Working Paper Series 1, 59. London: Crisis States Research Centre, London School of Economics and Political Science.

Beall, Jo. 2006. "Cultural Weapons: Traditions, Inventions and the Transition to Democratic Governance in Metropolitan Durban." *Urban Studies* 43, no. 2: 457–73.

Bechtel, Michael M., and Johannes Urpelainen. 2015. "All Policies Are Glocal: International Environmental Policy Making with Strategic Subnational Governments." *British Journal of Political Science* 45, no. 3: 559–82.

Beddington, John R., David J. Agnew, and Colin W. Clark. 2007. "Current Problems in the Management of Marine Fisheries." *Science* 316, no. 5832 (June 22): 1713–16.

Bednar, Jenna. 2006. "Is Full Compliance Possible? Conditions for Shirking with Imperfect Monitoring and Continuous Action Spaces." *Journal of Theoretical Politics* 18, no. 3: 347–75.

Bednar, Jenna, and Scott Page. 2007. "Can Game(s) Theory Explain Culture? The Emergence of Cultural Behavior within Multiple Games." *Rationality and Society* 19, no. 1: 65–97.

Bianco, William T., and Robert H. Bates. 1990. "Cooperation by Design: Leadership, Structure, and Collective Dilemmas." *American Political Science Review* 84, no. 1: 133–47.

Binmore, Ken. 2007a. *Game Theory: A Very Short Introduction.* Oxford: Oxford University Press.

Binmore, Ken. 2007b. *Playing for Real: A Text on Game Theory.* New York: Oxford University Press.

Boix, Carles.1999. "Setting the Rules of the Game: The Choice of Electoral Systems in Advanced Democracies." *American Political Science Review* 93, no. 3: 609–24.

Boix, Carles, and Susan C. Stokes. 2003. "Endogenous Democratization." *World Politics* 55, no. 4: 517–49.

Bollen, Kenneth, John T. Cacioppo, Robert M. Kaplan, Jon A. Krosnick, and James L. Olds. 2015. "Social, Behavioral, and Economic Sciences Perspectives on Robust and Reliable Science: Report of the Subcommittee on Replicability in Science, Advisory Committee to the National Science Foundation Directorate for Social, Behavioral, and Economic Sciences." Washington, DC: National Science Foundation. https://www.nsf.gov/sbe/AC_Materials/SBE_Robust_and_Reliable_Research_Report.pdf

Bolton, Patrick, and Mathias Dewatripont. 2005. *Contract Theory.* Cambridge: MIT Press.

Brams, Steven J. 2014. *Rational Politics: Decisions, Games, and Strategy.* San Diego: Academic Press.

Branch, Trevor A. 2009. "How Do Individual Transferable Quotas Affect Marine Ecosystems?" *Fish and Fisheries* 10, no. 1: 39–57.

Branch, Trevor A., and Danika Kleiber. 2015. "Should We Call Them Fishers or Fishermen?" *Fish and Fisheries* 18, no. 1 (September 23): 114–27.

Brander, James A., and Barbara J. Spencer. 1983. "Strategic Commitment with R&D: The Symmetric Case." *Bell Journal of Economics* 14, no. 1 (Spring): 225–35.

Brehm, John, and Scott Gates. 1999. *Working, Shirking, and Sabotage: Bureaucratic Response to a Democratic Public.* Ann Arbor: University of Michigan Press.

Buchanan, James M., and Gordon Tullock. 1962. *The Calculus of Consent.* Ann Arbor: University of Michigan Press.

Bueno de Mesquita, Bruce, et al. 2003. "Symposium on Replication in International Studies Research." *International Studies Perspectives* 4, no. 1: 72–107.

Camerer, Colin F., Anna Dreber, Eskil Forsell, Teck-Hua Ho, Jürgen Huber, Magnus Johannesson, Michael Kirchler, et al. 2016. "Evaluating Replicability of Laboratory Experiments in Economics." *Science* 351, no. 6280: 1433–36.

Carnap, Rudolf. 1936. "Testability and Meaning." *Philosophy of Science* 3, no. 4: 419–71.

Carnap, Rudolf. 1966. *Philosophical Foundations of Physics.* New York: Basic Books.

Carnap, Rudolf. 1987. "The Confirmation of Laws and Theories." In *Scientific Knowledge: Basic Issues in the Philosophy of Science*, edited by Janet A. Kourany. Belmont, CA: Wadsworth.

Carpenter, Daniel. 2000. "Commentary: What Is the Marginal Value of Analytic Narratives?" *Social Science History* 24, no. 4: 653–67.

Carsey, Thomas M. 2014. "Making DA-RT a Reality." *PS: Political Science & Politics* 47, no. 1: 72–77.

Cartwright, Nancy. 1999. *The Dappled World: A Study of the Boundaries of Science.* Cambridge: Cambridge University Press.

Chandra, Kanchan. 2001. "Ethnic Bargains, Group Instability, and Social Choice Theory." *Politics & Society* 29, no. 3: 337–62.

Cherry, Todd L., Jon Hovi, and David M. McEvoy, eds. 2014. *Toward a New Climate Agreement: Conflict, Resolution and Governance.* New York: Routledge.

Chin-Yee, Simon. 2016. "Briefing: Africa and the Paris Climate Change Agreement." *African Affairs* 115, no. 459: 359–68.

Christensen, Clayton M. 2006. "The Ongoing Process of Building a Theory of Disruption." *Journal of Product Innovation Management* 23, no. 1: 39–55.

Cingranelli, David, Paola Fajardo-Heyward, and Mikhail Filippov. 2014. "Principals, Agents and Human Rights." *British Journal of Political Science* 44, no. 3 (July): 605–30.

Cingranelli, David, and Mikhail Filippov. 2018a. "Are Human Rights Practices Improving?" *American Political Science Review* 112, no. 4: 1–7.

Cingranelli, David, and Mikhail Filippov. 2018b. "Problems of Model Specification and Improper Data Extrapolation." *British Journal of Political Science* 48, no. 1: 273–74.

Clarke, Kevin A., and David M. Primo. 2007. "Modernizing Political Science: A Model-Based Approach." *Perspectives on Politics* 5, no. 4: 741–53.

Clarke, Kevin A., and David M. Primo. 2012. *A Model Discipline: Political Science and the Logic of Representations.* Oxford: Oxford University Press.

Clemens, Michael A. 2017. "The Meaning of Failed Replications: A Review and Proposal." *Journal of Economic Surveys* 31, no. 1: 326–42.

Climate Action Tracker. 2018. *Climate Action Tracker: Countries, USA.* Accessed June 2018. https://climateactiontracker.org/countries/usa/

Coecke, Bob, and Sonja Smets. 2004. "The Sasaki Hook Is Not a [Static] Implicative Connective but Induces a Backward [in Time] Dynamic one that Assigns Causes." *International Journal of Theoretical Physics* 43, no. 7–8: 1705–36.

Condorcet, Marie Jean Antoine Nicolas de, and Marquis de Caritat. 1788. "On the Constitution and the Functions of Provincial Assemblies." Translated and excerpted in McLean Iain, and Fiona Hewitt, eds. 1994. *Condorcet: Foundations of Social Choice and Political Theory.* Cheltenham: Edward Elgar, 1994, 139–68.

Costello, Christopher, Steven D. Gaines, and John Lynham. 2008. "Can Catch Shares Prevent Fisheries Collapse?" *Science* 321, no. 5896 (September 19): 1678–81.

Cowen, Tyler. 1998. "Do Economists Use Social Mechanisms to Explain?" In *Social Mechanisms: An Analytical Approach to Social Theory*, edited by Peter Hedström, Richard Swedberg, and Gudmund Hernes, 125–46. New York: Cambridge University Press.

Crettez, Bertrand, and Régis Deloche. 2018. "An Analytic Narrative of Caesar's Death: Suicide or Not? That Is the Question." *Rationality and Society* 30, no. 3: 332–49.

Crouch, Ben M., and James W. Marquart. 1989. *An Appeal to Justice: Litigated Reform of Texas Prisons.* Austin: University of Texas Press.

Dafoe, Allan. 2014. "Science Deserves Better: The Imperative to Share Complete Replication Files." *PS: Political Science & Politics* 47, no. 1: 60–66.

de Kadt, Daniel, and Horacio A. Larreguy. 2018. "Agents of the Regime? Traditional Leaders and Electoral Politics in South Africa." *Journal of Politics* 80, no. 2: 382–99.

DeLong, J. Bradford. 2003. "India since Independence: An Analytic Growth Narrative." In *Search of Prosperity: Analytic Narratives on Economic Growth*, edited by Dani Rodrik, 184–204. Princeton: Princeton University Press.

Dewald, William G., Jerry G. Thursby, and Richard G. Anderson. 1986. "Replication in Empirical Economics: The Journal of Money, Credit and Banking Project." *American Economic Review* 76, no. 4: 587–603.

Dixit, Avinash. 1980. "The Role of Investment in Entry-Deterrence." *Economic Journal* 90, no. 357 (March): 95–106.

Dixit, Avinash K., and Barry J. Nalebuff. 1993. *Thinking Strategically: The Competitive Edge in Business, Politics, and Everyday Life*. New York: W. W. Norton.

Dixit, Avinash, David Reiley, and Susan Skeath. 2009. *Games of Strategy*. New York: W. W. Norton.

Downs, Anthony. 1957. "An Economic Theory of Political Action in a Democracy." *Journal of Political Economy* 65, no. 2: 135–50.

Edwards, Steven F. 2003. "Property Rights to Multi-Attribute Fishery Resources." *Ecological Economics* 44, nos. 2–3 (March): 309–23.

Elliott, Jane. 1999. "Models Are Stories Are Not Real Life." In *Statistics in Society: The Arithmetic of Politics*, edited by Daniel Dorling and Stephen Simpson, 95–102. London: Arnold.

Elliott, Jane. 2005. *Using Narrative in Social Research: Qualitative and Quantitative Approaches*. Thousand Oaks, CA: Sage.

Elman, Colin, and Diana Kapiszewski. 2014. "Data Access and Research Transparency in the Qualitative Tradition." *PS: Political Science & Politics* 47, no. 1: 43–47.

Elster, Jon. 1998. "A Plea for Mechanisms." In *Social Mechanisms: An Analytical Approach to Social Theory*, edited by Peter Hedström, Richard Swedberg, and Gudmund Hernes, 45–73. New York: Cambridge University Press.

Elster, Jon. 2000. "Rational Choice History: A Case of Excessive Ambition." *American Political Science Review* 94, no. 3: 685–95.

Elster, Jon. 2007. *Explaining Social Behavior: More Nuts and Bolts for the Social Sciences*. New York: Cambridge University Press.

Ensminger, Jean. 1996. *Making a Market: The Institutional Transformation of an African Society*. New York: Cambridge University Press.

Epstein, Lee, and Carol Mershon. 1996. "Measuring Political Preferences." *American Journal of Political Science* 40, no. 1: 261–94.

Epstein, Lee, and Olga Shvetsova. 2002. "Heresthetical Maneuvering on the US Supreme Court." *Journal of Theoretical Politics* 14, no. 1: 93–122.

Farrer, Benjamin, R. Holahan, and Olga Shvetsova. 2017. "Accounting for Heterogeneous Private Risks in the Provision of Collective Foods: Controversial Compulsory Contracting Institutions in Horizontal Hydrofracturing." *Journal of Economic Behavior & Organization* 133: 138–50.

Fearon, James D. 1994. "Domestic Political Audiences and the Escalation of International Disputes." *American Political Science Review* 88, no. 3 (September): 577–92.

Feddersen, Timothy J. 2004. "Rational Choice Theory and the Paradox of Not Voting." *Journal of Economic Perspectives* 18, no. 1: 99–112.

Federal Bureau of Investigation. 2017. "About: Frequently Asked Questions." April 28. Accessed June 2017. https://www.fbi.gov/about/faqs

Feeny, David, Fikret Berkes, Bonnie J. Mccay, and James M. Acheson. 1990. "The

Tragedy of the Commons: Twenty-Two Years Later." *Human Ecology* 18, no. 1 (March): 1–19.

Fellner, Jamie. 2015. "Callous and Cruel: Use of Force against Inmates with Mental Disabilities in US Jails and Prisons." Human Rights Watch. May 12. Accessed June 2017. https://www.hrw.org/report/2015/05/12/callous-and-cruel/use-force-against-inmates-mental-disabilities-us-jails-and

Ferejohn, John. 1986. "Incumbent Performance and Electoral Control." *Public Choice* 50, no. 1: 5–25.

Ferejohn, John, 2007. Positive Theory and the Internal View of Law. *University of Pennsylvania Journal of Constitutional Law* 10 (2): 273–303.

Ferejohn, John A., and Morris P. Fiorina. 1974. "The Paradox of Not Voting: A Decision Theoretic Analysis." *American Political Science Review* 68, no. 2: 525–36.

Filippov, Mikhail, Peter C. Ordeshook, and Olga Shvetsova. 2004. *Designing Federalism: A Theory of Self-Sustainable Federal Institutions*. Cambridge: Cambridge University Press.

Fink, Evelyn C., Scott Gates, and Brian D. Humes. 1998. *Game Theory Topics: Incomplete Information, Repeated Games, and N-Player Games*. No. 122. Thousand Oaks: Sage.

Fiorina, Morris P. 1975. "Formal Models in Political Science." *American Journal of Political Science* 19, no. 1: 133–59.

Fishburn, Peter C. 2015. *The Theory of Social Choice*. Princeton: Princeton University Press.

Flach, Peter A., and Antonis Hadjiantonis, eds. 2013. *Abduction and Induction: Essays on Their Relation and Integration*. Berlin: Springer Science & Business Media.

Flach, Peter A., and Antonis C. Kakas. 2000. "Abductive and Inductive Reasoning: Background and Issues." In *Abduction and Induction*, edited by Peter A. Flach and Antonis Hadjiantonis, 1–27. Dordrecht: Springer.

Flach, Peter, Antonis Kakas, and Oliver Ray. 2006. "Abduction, Induction, and the Logic of Scientific Knowledge Development." In *Workshop on Abduction and Induction in AI and Scientific Modelling, Program chairs Peter Flach, Antonis Kakas, Lorenzo Magnani, and Oliver Ray, 21-3. Riva del Garda, August*. https://www.researchgate.net/profile/Antonis_Kakas/publication/242402242_Workshop_on_Abduction_and_Induction_in_AI_and_Scientific_Modelling/links/004635321627f3fb76000000.pdf#page=25

Food and Agriculture Organization of the UN. 2014. *FAO Yearbook. Fishery and Aquaculture Statistics Report*. Rome: FAO. http://www.fao.org/fishery/static/Yearbook/YB2014_CD_Master/navigation/index_intro_e.htm

Freese, Jeremy. 2007. "Replication Standards for Quantitative Social Science: Why Not Sociology?" *Sociological Methods & Research* 36, no. 2: 153–72.

Fuess, Scott M., Jr. 1996. "On Replication in Business and Economics Research: The QJBE Case." *Quarterly Journal of Business and Economics* 35, no. 2: 3–13.

Gartner, Scott Sigmund, and Patrick M. Regan. 1996. "Threat and Repression: The Non-linear Relationship between Government and Opposition Violence." *Journal of Peace Research* 33, no. 3: 273–87.

Gates, Scott, and Brian D. Humes. 1997. *Games, Information, and Politics: Applying Game Theoretic Models to Political Science*. Ann Arbor: University of Michigan Press.

Gehlbach, Scott. 2013. *Formal Models of Domestic Politics*. New York: Cambridge University Press.

George, Alexander L., and Andrew Bennett. 2005. *Case Studies and Theory Development in the Social Sciences*. Cambridge: MIT Press.

Gerber, Elisabeth R. 1996. "Legislative Response to the Threat of Popular Initiatives." *American Journal of Political Science* 40, no. 1: 99–128.

Gibbard, Allan, and Hal R. Varian. 1978. "Economic Models." *Journal of Philosophy* 75, no. 11: 664–77.

Gibbs, Mark T. 2009. "Individual Transferable Quotas and Ecosystem-Based Fisheries Management: It's All in the T." *Fish and Fisheries* 10, no. 4 (September 25): 470–74.

Gilbert, Daniel T., Gary King, Stephen Pettigrew, and Timothy D. Wilson. 2016. "Comment on 'Estimating the Reproducibility of Psychological Science.'" *Science* 351, no. 6277: 1037.

Gintis, Herbert. 2009. *Game Theory Evolving: A Problem-Centered Introduction to Modeling Strategic Interaction*. Princeton: Princeton University Press.

Gómez, Omar S., Natalia Juristo, and Sira Vegas. 2014. "Understanding Replication of Experiments in Software Engineering: A Classification." *Information and Software Technology* 56, no. 8: 1033–48.

Gottschalk, Marie. 2008. "Hiding in Plain Sight: American Politics and the Carceral State." *Annual Review of Political Science* 11: 235–60.

Grafton, R. Q., Ragnar Arnason, Trond Bjørndal, David Campbell, Harry F. Campbell, Colin W. Clark, Robin Connor, et al. 2006. "Incentive-Based Approaches to Sustainable Fisheries." *Canadian Journal of Fisheries and Aquatic Sciences* 63 (February 15): 699–710.

Granato, Jim, and Frank Scioli. 2004. "Puzzles, Proverbs, and Omega Matrices: The Scientific and Social Significance of Empirical Implications of Theoretical Models (EITM)." *Perspectives on Politics* 2, no. 2: 313–23.

Green, Donald, Bradley Palmquist, and Eric Schickler. 1998. "Macropartisanship: A Replication and Critique." *American Political Science Review* 92, no. 4: 883–99.

Green, Donald, and Ian Shapiro. 1994. *Pathologies of Rational Choice Theory: A Critique of Applications in Political Science*. New Haven: Yale University Press.

Groseclose, Tim, and Nolan McCarty. 2001. "The Politics of Blame: Bargaining before an Audience." *American Journal of Political Science* 45, no. 1: 100–119.

Grossman, Gene M., and Elhanan Helpman. 2001. *Special Interest Politics.* Cambridge: MIT Press.

Hamermesh, Daniel S. 2007. "Replication in Economics." *Canadian Journal of Economics/Revue canadienne d'économique* 40, no. 3: 715–33.

Hands, D. Wade. 1991. "Popper, the Rationality Principle, and Economic Explanation." In *Economics, Culture, and Education: Essays in Honor of Mark Blaug,* edited by G. K. Shaw, 114–15. Cheltenham, UK: Edward Elgar.

Hardin, Garrett. 1968. "The Tragedy of the Commons." *Science* 162, no. 3859: 1243–48.

Hardin, Garrett. 1998. "Extensions of 'The Tragedy of the Commons.'" *Science* 280, no. 5364 (May 1): 682–83.

Hardin, Russell. 1971. "Collective Action as an Agreeable N-Prisoners' Dilemma." *Behavioral Science* 16, no. 5: 472–81.

Hardin, Russell. 1982. *Collective Action.* Baltimore: Johns Hopkins University Press.

Hartzell, Caroline. 2015. "Data Access and Research Transparency (DA-RT): A Joint Statement by Political Science Journal Editors." *Conflict Management and Peace Science* 32, no. 4: 355.

Hausman, Daniel M. 2009. "Laws, Causation, and Economic Methodology." In *The Oxford Handbook of Philosophy of Economics,* edited by Don Ross and Harold Kincaid, 35–54. New York: Oxford University Press.

Hawken, Paul. 2017. *Drawdown: The Most Comprehensive Plan Ever Proposed to Reverse Global Warming.* New York: Penguin Books.

Hedley, Chris. 2001. "The 1998 Agreement on the International Dolphin Conservation Program: Recent Developments in the Tuna-Dolphin Controversy in the Eastern Pacific Ocean." *Ocean Development & International Law* 32, no. 1: 71–92.

Heller, William B. 1997. "Bicameralism and Budget Deficits: The Effect of Parliamentary Structure on Government Spending." *Legislative Studies Quarterly* 22, no. 4: 485–516.

Heller, William B. 2001a. "Making Policy Stick: Why the Government Gets What It Wants in Multiparty Parliaments." *American Journal of Political Science* 45, no. 4: 780–98.

Heller, William B. 2001b. "Political Denials: The Policy Effect of Intercameral Partisan Differences in Bicameral Parliamentary Systems." *Journal of Law, Economics, and Organization* 17, no. 1: 34–61.

Heller, William B. 2002. "Regional Parties and National Politics in Europe: Spain's Estado de las Autonomías, 1993 to 2000." *Comparative Political Studies* 35, no. 6: 657–85.

Heller, William B., and Carol Mershon. 2008. "Dealing in Discipline: Party Switching and Legislative Voting in the Italian Chamber of Deputies, 1988–2000." *American Journal of Political Science* 52, no. 4: 910–25.

Heller, William B., and Olga Shvetsova. 2016. "The Ouroborous of Political Institutions: Party Rules in Institutional Context." Citizenship, Rights, and Cultural Belonging Working Paper No. 101, Binghamton University, October.

Heller, William B., and Katri K. Sieberg. 2008. "Functional Unpleasantness: The Evolutionary Logic of Righteous Resentment." *Public Choice* 135, no. 3: 399–413.

Heller, William B., and Katri K. Sieberg. 2010. "Honor among Thieves: Cooperation as a Strategic Response to Functional Unpleasantness." *European Journal of Political Economy* 26, no. 3: 351–62.

Hempel, Carl G. 1965. *Aspects of Scientific Explanation and Other Essays in the Philosophy of Science*. New York: Free Press.

Hillers, Mike. 2016. "Trawling in the Age of Technology." *Fishermen's News*, August 1. Accessed June 2017. http://www.fishermensnews.com/story/2016/08/01/features/trawling-in-the-age-of-technology/409.html

Hinchman, Lewis P., and Sandra Hinchman, eds. 1997. *Memory, Identity, Community: The Idea of Narrative in the Human Sciences*. Albany: SUNY Press.

Hovi, Jon, Hugh Ward, and Frank Grundig. 2015. "Hope or Despair? Formal Models of Climate Cooperation." *Environmental and Resource Economics* 62, no. 4: 665–88.

Hovi, Jon, Detief F. Sprinz, Håkon Sælen, and Arild Underdal. 2016. "Climate Change Mitigation: A Role for Climate Clubs?" *Palgrave Communications* 2: no. 1: 1–9.

Htun, Mala. 2016. "DA-RT and the Social Conditions of Knowledge Production in Political Science." *Comparative Politics Newsletter* 26, no. 1: 32–35.

Hubbard, Raymond, and Daniel E. Vetter. 1996. "An Empirical Comparison of Published Replication Research in Accounting, Economics, Finance, Management, and Marketing." *Journal of Business Research* 35, no. 2: 153–64.

Huber, John D., and Arthur Lupia. 2001. "Cabinet Instability and Delegation in Parliamentary Democracies." *American Journal of Political Science* 45, no. 1: 18–32.

Huber, John D., and Nolan McCarty. 2004. "Bureaucratic Capacity, Delegation, and Political Reform." *American Political Science Review* 98, no. 3: 481–94.

Humphreys, Macartan. 2016. *Political Games: Mathematical Insights on Fighting, Voting, Lying, & Other Affairs of State*. New York: W. W. Norton.

Hunt, Karl. 1975. "Do We Really Need More Replications?" *Psychological Reports* 36, no. 2: 587–93.

Hyman, Herbert H., and Charles R. Wright. 1967. "Evaluating Social Action Programs." In *The Uses of Sociology*, edited by Paul F. Lazarsfeld, et al., 741–82. New York: Basic Books.

Hymes, Dell. 2003. *Ethnography, Linguistics, Narrative Inequality: Toward an Understanding of Voice*. Abingdon: Taylor and Francis.

Hyvärinen, Matti. 2016. "Narrative and Sociology." *Narrative Works* 6, no. 1: 38–62.

IPCC (Intergovernmental Panel on Climate Change). 2018. *Global Warming of 1.5°C. An IPCC Special Report on the Impacts of Global Warming of 1.5°C above Pre-Industrial Levels and Related Global Greenhouse Gas Emission Pathways in the Context of Strengthening the Global Response to the Threat of Climate Change, Sustainable Development, and Efforts to Eradicate Poverty* [Masson-Delmotte, V., P. Zhai, H. O. Pörtner, D. Roberts, J. Skea, P.R. Shukla, et al., eds.]. Geneva: Intergovernmental Panel on Climate Change. In Press.

Isaac, Jeffrey C. 2015. "For a More Public Political Science." *Perspectives on Politics* 13, no. 2: 269–83.

Ishiyama, J., 2014. "Replication, Research Transparency, and Journal Publications: Individualism, Community Models, and the Future of Replication Studies." *PS: Political Science & Politics* 47, no. 1: 78–83.

Johnson, James. 2012. "What Rationality Assumption? Or, How 'Positive Political Theory' Rests on a Mistake." *Political Studies* 58 (2): 282–99.

Kalyvas, Stathis N. 1996. *The Rise of Christian Democracy in Europe*. New York: Cornell University Press.

Kane, Edward J. 1984. "Why Journal Editors Should Encourage the Replication of Applied Econometric Research." *Quarterly Journal of Business and Economics* 23, no. 1: 3–8.

Keohane, Robert O. 2015. "The Global Politics of Climate Change: Challenge for Political Science." *PS: Political Science & Politics* 48, no. 1: 19–26.

Kerin, Roger A., P. Rajan Varadarajan, and Robert A. Peterson. 1992. "First-Mover Advantage: A Synthesis, Conceptual Framework, and Research Propositions." *Journal of Marketing* 56, no. 4 (October): 33–52.

Kiewiet, D. Roderick, and Mathew Daniel McCubbins. 1991. *The Logic of Delegation: Congressional Parties and the Appropriations Process*. Chicago: University of Chicago Press.

Kim, Victoria. 2014. "Six L.A. County Sheriff Workers Get Prison for Obstructing Jail Probe." *Los Angeles Times*, September 23. Accessed May 2017. http://www.latimes.com/local/countygovernment/la-me-deputy-corruption-20140924-story.html

King, Gary. 1995. "Replication, Replication." *PS: Political Science & Politics* 28, no. 3: 444–52.

King, Gary. 2011. "Ensuring the Data-Rich Future of the Social Sciences." *Science* 331, no. 6018: 719–21.

Kinsey, Barbara Sgouraki, and Olga Shvetsova. 2008. "Applying the Methodology of Mechanism Design to the Choice of Electoral Systems." *Journal of Theoretical Politics* 20, no. 3: 303–27.

Kirkley, James, Dale Squires, and Ivar E. Strand. 1998. "Characterizing Managerial Skill and Technical Efficiency in a Fishery." *Journal of Productivity Analysis* 9, no. 2: 145–60.

Klein, Naomi. 2014. *This Changes Everything: Capitalism versus the Climate*. New York: Simon & Schuster.

Kollman, Ken, John H. Miller, and Scott E. Page. 1992. "Adaptive Parties in Spatial Elections." *American Political Science Review* 86, no. 4: 929–37.

Kornblum, Allan N. 1976. *The Moral Hazards: Police Strategies for Honesty and Ethical Behavior*. Lexington, MA: D. C. Heath.

Koter, Dominika. 2013. "King Makers: Local Leaders and Ethnic Politics in Africa." *World Politics* 65, no. 2: 187–232.

Kramer, Gerald H. 1986. "Political Science as Science." In *Political Science: The Science of Politics*, edited by Herbert F. Weisberg, 11–23. New York: Agathon Press.

Kuhn, Thomas S. 1962. *The Structure of Scientific Revolutions*. Chicago: University of Chicago Press.

Labov, William. 2003. "Uncovering the Event Structure of Narrative." *Georgetown University Roundtable on Languages and Linguistics (GURT) 2001*. Washington, DC: Georgetown University Press, 63–83.

Labov, William. 2006. "Narrative Pre-construction." *Narrative Inquiry* 16, no. 1: 37–45.

Lakatos, Imre, and Alan Musgrave, eds. 1970. *Criticism and the Growth of Knowledge: Proceedings of the International Colloquium in the Philosophy of Science*. Cambridge: Cambridge University Press.

La Sorte, Michael A. 1972. "Replication as a Verification Technique in Survey Research: A Paradigm." *Sociological Quarterly* 13, no. 2: 218–27.

Laver, Michael, and Norman Schofield. 1990. *Multiparty Government: The Politics of Coalition in Europe*. New York: Oxford University Press.

Laver, Michael, and Kenneth A. Shepsle. 1990. "Coalitions and Cabinet Government." *American Political Science Review* 84, no. 3: 873–90.

Laver, Michael, and Kenneth A. Shepsle. 1996. *Making and Breaking Governments: Cabinets and Legislatures in Parliamentary Democracies*. Cambridge: Cambridge University Press.

Laver, Michael, and Kenneth A. Shepsle. 1998. "Events, Equilibria, and Government Survival." *American Journal of Political Science* 42, no. 1: 28–54.

Lerman, Amy E., and Vesla M. Weaver. 2014. *Arresting Citizenship: The Democratic Consequences of American Crime Control*. Chicago: University of Chicago Press.

Levi, Margaret. 2002. "Modeling Complex Historical Processes with Analytic Narratives." In *Akteure, Mechanismen, Modelle. Zur Theoriefähigkeit makro-sozialer Analysen*, edited by R. Mayntz, 108–27. Frankfurt: Campus Verlag.

Levi, Margaret. 2004. "An Analytic Narrative Approach to Puzzles and Problems." In *Problems and Methods in the Study of Politics*, edited by I. Shapiro, R. Smith, and T. Masoud, 201–26. Cambridge: Cambridge University Press.

Lewin, Leif. 1988. *Ideology and Strategy: A Century of Swedish Politics*. New York: Cambridge University Press.

Lieberman, Evan S. 2010. "Bridging the Qualitative-Quantitative Divide: Best Practices in the Development of Historically Oriented Replication Databases." *Annual Review of Political Science* 13: 37–59.

Logan, Carolyn. 2009. "Selected Chiefs, Elected Councillors and Hybrid Democrats: Popular Perspectives on the Co-existence of Democracy and Traditional Authority." *Journal of Modern African Studies* 47, no. 1: 101–28.

Lupia, Arthur. 2008. "Procedural Transparency and the Credibility of Election Surveys." *Electoral Studies* 27, no. 4: 732–39.

Lupia, Arthur, and George Alter. 2014. "Data Access and Research Transparency in the Quantitative Tradition." *PS: Political Science & Politics* 47, no. 1: 54–59.

Lupia, Arthur, and Colin Elman. 2014. "Openness in Political Science: Data Access and Research Transparency." *PS: Political Science & Politics* 47, no. 1: 19–42.

Lupia, Arthur, and Kaare Strøm. 1995. "Coalition Termination and the Strategic Timing of Parliamentary Elections." *American Political Science Review* 89, no. 3: 648–65.

MacDonald, Paul K. 2003. "Useful Fiction or Miracle Maker: The Competing Epistemological Foundations of Rational Choice Theory." *American Political Science Review* 97, no. 4: 551–65.

Machiavelli, Niccolò. 1975. *The Prince*. Translated by George Bull. London: Penguin Books.

Madani, Kaveh. 2013. "Modeling International Climate Change Negotiations More Responsibly: Can Highly Simplified Game Theory Models Provide Reliable Policy Insights?" *Ecological Economics* 90: 68–76.

Maines, David R. 1993. "Narrative's Moment and Sociology's Phenomena: Toward a Narrative Sociology." *The Sociological Quarterly* 34, no. 1: 17–38.

Mares, Isabela, and Lauren Young. 2016. "Buying, Expropriating, and Stealing Votes." *Annual Review of Political Science* 19: 267–88.

Martin, Michael W., and Jane Sell. 1979. "The Role of the Experiment in the Social Sciences." *Sociological Quarterly* 20, no. 4: 581–90.

McCarty, Nolan, and Adam Meirowitz. 2007. *Political Game Theory: An Introduction*. New York: Cambridge University Press.

McGuire, Michael, and Jenna Bednar. 2010. "The Robust Federation: Principles of Design." *Perspectives on Politics* 8, no. 1: 373.

McKelvey, Richard D. 1976. "Intransitivities in Multidimensional Voting Models and Some Implications for Agenda Control." *Journal of Economic Theory* 12, no. 3: 472–82.

McKelvey, Richard D. 1979. "General Conditions for Global Intransitivities in Formal Voting Models." *Econometrica: Journal of the Econometric Society* 47, no. 5: 1085–1112.

McKelvey, Richard D., and Norman Schofield. 1986. "Structural Instability of the Core." *Journal of Mathematical Economics* 15, no. 3: 179–98.

McKelvey, Richard D., and Norman Schofield. 1987. "Generalized Symmetry Conditions at a Core Point." *Econometrica: Journal of the Econometric Society* 55, no. 4: 923–33.

McLean, Iain, and Arnold B. Urken. 1995. *Classics of Social Choice*. Ann Arbor: University of Michigan Press.

McNeeley, Susan, and Jessica J. Warner. 2015. "Replication in Criminology: A Necessary Practice." *European Journal of Criminology* 12, no. 5: 581–97.

Medina, Luis Fernando. 2013. "The Analytical Foundations of Collective Action Theory: A Survey of Some Recent Developments." *Annual Review of Political Science* 16: 259–83.

Meng, Anne. 2018a. "Accessing the State: Executive Constraints and Credible Commitment in Dictatorships." Working Paper. University of Virginia.

Meng, Anne. 2018b. "Tying the Big Man's Hands: From Personalized Rule to Institutionalized Regimes." Book manuscript. University of Virginia.

Mershon, Carol. 2002. *The Costs of Coalition*. Stanford: Stanford University Press.

Mershon, Carol. 2017. "Local Public Goods and Political Elites: Elective and Chiefly Authority in South Africa." Paper presented at the 2017 Meetings of the Midwest Political Science Association, Chicago, April 6–9.

Mershon, Carol, and Olga Shvetsova. 2011. "Moving in Time: Legislative Party Switching as Time-Contingent Choice." In *Political Economy of Institutions, Democracy and Voting*, edited by Norman Schofield and Gonzalo Caballero, 389–402. Berlin: Springer.

Mershon, Carol, and Olga Shvetsova. 2013a. "The Microfoundations of Party System Stability in Legislatures." *Journal of Politics* 75, no. 4: 865–78.

Mershon, Carol, and Olga V. Shvetsova. 2013b. *Party System Change in Legislatures Worldwide: Moving outside the Electoral Arena*. New York: Cambridge University Press.

Mershon, Carol, and Olga Shvetsova. 2014. "Change in Parliamentary Party Systems and Policy Outcomes: Hunting the Core." *Journal of Theoretical Politics* 26, no. 2: 331–51.

Mershon, Carol, and Olga Shvetsova. 2018. "Meta-Constitutional Bargaining for Legitimacy in Dual Legitimacy Systems." Paper presented at the 2018 Meetings of the American Political Science Association, Boston, August 30–September 2.

Mershon, Carol, and Olga Shvetsova. 2019. "Traditional Authority and Bargaining for Legitimacy in Dual Legitimacy Systems." Forthcoming, *Journal of Modern African Studies.*

Miller, Gary, and Norman Schofield. 2003. "Activists and Partisan Realignment in the United States." *American Political Science Review* 97, no. 2: 245–60.

Monbiot, George. 2006. "Drastic Action on Climate Change Is Needed Now—and Here's the Plan." *Guardian,* October 31, 2006. https://www.theguardian.com/commentisfree/2006/oct/31/economy.politics

Monbiot, George. 2007. *Heat: How to Stop the Planet from Burning.* Cambridge, MA: South End Press.

Monroe, Nathan W., and Gregory Robinson. 2008. "Do Restrictive Rules Produce Nonmedian Outcomes? A Theory with Evidence from the 101st–108th Congresses." *Journal of Politics* 70, no. 1: 217–31.

Morrow, James D. 1994. *Game Theory for Political Scientists.* Princeton: Princeton University Press.

Morton, Rebecca B. 1999. *Methods and Models: A Guide to the Empirical Analysis of Formal Models in Political Science.* New York: Cambridge University Press.

Munasinghe, Mohan, and Rob Swart. 2005. *Primer on Climate Change and Sustainable Development: Facts, Policy Analysis, and Applications.* Cambridge: Cambridge University Press.

Myerson, Roger B. 2013. *Game Theory: Analysis of Conflict.* Cambridge: Harvard University Press.

Nalepa, Monika. 2010. "Captured Commitments: An Analytic Narrative of Transitions with Transitional Justice." *World Politics* 62, no. 2: 341–80.

Nash, Cristopher, ed. 2005. *Narrative in Culture: The Uses of Storytelling in the Sciences, Philosophy and Literature.* New York: Routledge.

National Center for Victims of Crime. 2019. "The Criminal Justice System." Accessed February 2019. http://victimsofcrime.org/help-for-crime-victims/get-help-bulletins-for-crime-victims/the-criminal-justice-system

National Science Foundation. 2016. "Dear Colleague Letter: Robust and Reliable Research in the Social, Behavioral, and Economic Sciences." September 20. Accessed August 2017. https://www.nsf.gov/pubs/2016/nsf16137/nsf16137.jsp

Neely, Francis. 2007. "Party Identification in Emotional and Political Context: A Replication." *Political Psychology* 28, no. 6: 667–88.

Niou, Emerson, and Peter C. Ordeshook. 2015. *Strategy and Politics: An Introduction to Game Theory.* New York: Routledge.

North Pacific Fishery Management Council. 2017. "North Pacific Fishery Management Council." Accessed April. https://www.npfmc.org

Nosek, Brian A., George Alter, George C. Banks, Denny Borsboom, Sara D. Bowman, Steven J. Breckler, et al. 2015. "Promoting an Open Research Culture." *Science* 348, no. 6242: 1422–25.

Nosek, Brian A., George Alter, George Banks, Denny Borsboom, Sara Bowman, Steven Breckler, Stuart Buck, et al. 2016. "Transparency and Openness Promotion (TOP) Guidelines." OSF Preprints. https://osf.io/vj54c/

Ntsebeza, Lungisile. 2005. *Democracy Compromised: Chiefs and the Politics of the Land in South Africa*. Leiden: Brill.

O'Hara, Patrick. 2005. *Why Law Enforcement Organizations Fail: Mapping the Organizational Fault Lines in Policing*. Durham, NC: Carolina Academic Press.

Olson, Mancur. 1965. *Logic of Collective Action: Public Goods and the Theory of Groups*. Cambridge: Harvard University Press.

Olson, Mancur. 1993. "Dictatorship, Democracy, and Development." *American Political Science Review* 87, no. 3: 567–76.

Olson, Mancur. 2000. *Power and Prosperity: Outgrowing Communist and Capitalist Dictatorships*. New York: Basic Books.

Olson, Mancur. 1965. *The Logic of Collective Action*. Cambridge: Harvard University Press.

Oomen, Barbara. 2005. *Chiefs in South Africa: Law, Power and Culture in the Post-Apartheid Era*. Pietermar*itzburg: University of KwaZulu-Natal Press.

Ordeshook, Peter C. 1986. *Game Theory and Political Theory: An Introduction*. Cambridge: Cambridge University Press.

Osborne, Martin J. 2004. *An Introduction to Game Theory*. New York: Oxford University Press.

Ostrom, Elinor. 1990. *Governing the Commons: The Evolution of Institutions for Collective Action*. New York: Cambridge University Press.

Ostrom, Elinor. 1996. "Crossing the Great Divide: Coproduction, Synergy, and Development." *World Development* 24, no. 6: 1073–87.

Page, Scott E. 1999. "On the Emergence of Cities." *Journal of Urban Economics* 45, no. 1: 184–208.

Page, Scott E. 2018. *The Model Thinker: What You Need to Know to Make Data Work for You*. New York: Basic Books.

Palfrey, Thomas R., and Howard Rosenthal. 1985. "Voter Participation and Strategic Uncertainty." *American Political Science Review* 79, no. 1: 62–78.

Parkhe, Arvind. 1993. "'Messy' Research, Methodological Predispositions, and Theory Development in International Joint Ventures." *Academy of Management Review* 18, no. 2: 227–68.

Patterson, Molly, and Kristen Renwick Monroe. 1998. "Narrative in Political Science." *Annual Review of Political Science* 1: 315–31.

Pedriana, Nicholas. 2005. "Rational Choice, Structural Context, and Increasing Returns: A Strategy for Analytic Narrative in Historical Sociology." *Sociological Methods and Research* 33, no. 3: 349–82.

Péreau, J.-C., L. Doyen, L. R. Little, and O. Thébaud. 2012. "The Triple Bottom Line: Meeting Ecological, Economic and Social Goals with Individual Transferable Quotas." *Journal of Environmental Economics and Management* 63, no. 3 (May): 419–34.

Perry, Chad. 1998. "Processes of a Case Study Methodology for Postgraduate Research in Marketing." *European Journal of Marketing* 32, nos. 9–10: 785–802.

Petersilia, Joan. 2014. "California Prison Downsizing and Its Impact on Local Criminal Justice Systems." *Harvard Law & Policy Review* 8: 327–58.

Pikitch, E. K., et al. 2004. "Ecosystem-Based Fishery Management." *Science* 305, no. 5682: 346–47.

Pikitch, E. K., et al. 2005. "Status, Trends and Management of Sturgeon and Paddlefish Fisheries." *Fish and Fisheries* 6, no. 3: 233–65.

Plott, Charles R. 1967. "A Notion of Equilibrium and Its Possibility under Majority Rule." *American Economic Review* 57, no. 4: 787–806.

Plott, Charles R. 1976. "Axiomatic Social Choice Theory: An Overview and Interpretation." *American Journal of Political Science* 20, no. 3: 511–96.

Popper, Karl. 1972. *Objective Knowledge: An Evolutionary Approach*. New York: Oxford University Press.

Popper, Karl R.1978. *The Three Worlds: The Tanner Lecture on Human Values*. Ann Arbor: University of Michigan Press.

Powell, Robert. 1987. "Crisis Bargaining, Escalation, and MAD." *American Political Science Review* 81, no. 3: 717–35.

Powell, Robert. 1990. *Nuclear Deterrence Theory: The Search for Credibility*. Cambridge: Cambridge University Press.

Powell, Robert. 1999. *In the Shadow of Power: States and Strategies in International Politics*. Princeton: Princeton University Press.

Putnam, Hilary. 1991. "The 'Corroboration' of Theories." In *The Philosophy of Science*, edited by Richard Boyd, Philip Gasper, and John D. Trout, 121–37. Cambridge: MIT Press.

Putnam, Robert D. 1988. "Diplomacy and Domestic Politics: The Logic of Two-Level Games." *International Organization* 42, no. 3: 427–60.

Rasmusen, Eric. 2006. *Games and Information: An Introduction to Game Theory*. 4th Ed. Cambridge, MA: Wiley-Blackwell.

Reck, Gregory G. 1983. "Narrative Anthropology." *Anthropology and Humanism Quarterly* 8, no. 1: 8–12.

Richardson, Laurel. 1990. "Narrative and Sociology." *Journal of Contemporary Ethnography* 19, no. 1: 116–35.

Riker, William H. 1982. *Liberalism against Populism: A Confrontation between the Theory of Democracy and the Theory of Social Choice*. San Francisco: Freeman.

Riker, William H. 1986. *The Art of Political Manipulation*. New Haven: Yale University Press.

Rodrik, Dani. 2003. *In Search of Prosperity: Analytic Narratives on Economic Growth*. Princeton: Princeton University Press.

Rosenberg, Dina, and Olga Shvetsova. 2016. "Autocratic Health versus Democratic

Health: Different Outcome Variables for Health as a Factor versus Health as a Right." In *The Political Economy of Social Choices: Studies in Political Economy* edited by Maria Gallego and Norman Schofield, 1–20. Berlin: Springer.

Rubinstein, Ariel. 1991. "Comments on the Interpretation of Game Theory." *Econometrica* 59, no. 4: 909–24.

Saari, Donald G. 2012. *Geometry of Voting*. Berlin: Springer Science & Business Media.

Saari, Donald G., and Katri K. Sieberg. 2004. "Are Partwise Comparisons Reliable?" *Research in Engineering Design* 15, no. 1: 62–71.

Salanié, Bernard. 1997. *The Economics of Contracts: A Primer*. Cambridge: MIT Press.

Scheer, Roddy, and Doug Moss. 2017. "In a Scrape: Seafloor Trawling Threatens Deep Ocean Species." *Scientific American*. Accessed June 2017. https://www.scientificamerican.com/article/saving-deep-ocean-species/

Schelling, Thomas C. 1960. *The Strategy of Conflict*. Cambridge: Harvard University Press.

Schelling, Thomas C. 1966. *Arms and Influence*. New Haven: Yale University Press.

Schiemann, John W. 2007. "Bizarre Beliefs and Rational Choices: A Behavioral Approach to Analytic Narratives." *Journal of Politics* 69, no. 2: 511–24.

Schofield, Norman. 1983. "Generic instability of majority rule." *Review of Economic Studies* 50, no. 4: 695–705.

Schofield, Norman. 1986. "Existence of a 'Structurally Stable' Equilibrium for a Non-Collegial Voting Rule." *Public Choice* 51, no. 3: 267–84.

Schofield, Norman. 1993. "Political Competition and Multiparty Coalition Governments." *European Journal of Political Research* 23, no. 1: 1–33.

Schofield, Norman. 1999. "The Heart of the Atlantic Constitution: International Economic Stability, 1919–1998." *Politics & Society* 27, no. 2: 173–215.

Schofield, Norman. 2002. "Evolution of the Constitution." *British Journal of Political Science* 32, no. 1: 1–20.

Schofield, Norman. 2003. "Valence Competition in the Spatial Stochastic Model." *Journal of Theoretical Politics* 15, no. 4: 371–83.

Schofield, Norman. 2004. "Equilibrium in the Spatial 'Valence' Model of Politics." *Journal of Theoretical Politics* 16, no. 4: 447–81.

Schofield, Norman. 2006. *Architects of Political Change: Constitutional Quandaries and Social Choice Theory*. Cambridge: Cambridge University Press.

Schofield, Norman, Bernard Grofman, and Scott L. Feld. 1988. "The Core and the Stability of Group Choice in Spatial Voting Games." *American Political Science Review* 82, no. 1: 195–211.

Schofield, Norman, and Michael Laver. 1985. "Bargaining Theory and Portfolio Payoffs in European Coalition Governments 1945–83." *British Journal of Political Science* 15, no. 2: 143–64.

Schofield, Norman, and Itai Sened. 2006. *Multiparty Democracy: Elections and Legislative Politics.* New York: Cambridge University Press.

Schofield, Norman, Itai Sened, and David Nixon. 1998. "Nash Equilibrium in Multiparty Competition with 'Stochastic' Voters." *Annals of Operations Research* 84: 3–27.

Schwartz-Shea, Peregrine, and Dvora Yanow. 2016. "Legitimizing Political Science or Splitting the Discipline? Reflections on DA-RT and the Policy-Making Role of a Professional Association." *Politics & Gender* 12, no. 3.

Selten, Reinhard. 1978. "The Chain Store Paradox." *Theory and Decision* 9, no. 2 (April): 127–59.

Sen, Amartya. 1986. "Social Choice Theory." In *Handbook of Mathematical Economics*, vol. 3, edited by Michael Intriligator and Kenneth Arrow, 1073–1181. New York: Elsevier.

Shapiro, David. 2011. *Banking on Bondage: Private Prisons and Mass Incarceration.* New York: American Civil Liberties Union.

Shepsle, Kenneth A. 1979. "Institutional Arrangements and Equilibrium in Multidimensional Voting Models." *American Journal of Political Science* 23, no. 1: 27–59.

Shepsle, Kenneth A., and Mark S. Bonchek. 1997. *Analyzing Politics: Rationality, Behavior, and Institutions.* New York: W. W Norton.

Shepsle, Kenneth A., and Barry R. Weingast. 1981."Structure-Induced Equilibrium and Legislative Choice." *Public Choice* 37, no. 3: 503–19.

Shepsle, Kenneth A., and Barry R. Weingast. 1982. "Institutionalizing Majority Rule: A Social Choice Theory with Policy Implications." *American Economic Review* 72, no. 2: 367–71.

Shepsle, Kenneth A., and Barry R. Weingast. 1987. "The Institutional Foundations of Committee Power." *American Political Science Review* 81, no. 1: 85–104.

Sieberg, Katri K. 2005. *Criminal Dilemmas: Understanding and Preventing Crime.* 2nd ed. New York: Springer-Verlag Berlin Heidelberg.

Silberman, Matthew. 1995. *A World of Violence: Corrections in America.* Belmont, CA: Wadsworth.

Skultans, Vieda. 2008. *Empathy and Healing: Essays in Medical and Narrative Anthropology.* Oxford, NY: Berghahn Books.

Snyder, Glenn H. 1971. "'Prisoner's Dilemma' and 'Chicken' Models in International Politics." *International Studies Quarterly* 15, no. 1: 66–103.

Snyder, James M., Jr., and Michael M. Ting. 2002. "An Informational Rationale for Political Parties." *American Journal of Political Science* 46, no. 1: 90–110.

Soufiani, Hossein Azari, David C. Parkes, and Lirong Xia. 2014. "A Statistical Decision-Theoretic Framework for Social Choice." In *Advances in Neural Infor-*

mation Processing Systems, edited by David S. Touretzky, Michael C. Mozer, and Michael E. Hasselmo, 3185–93. Cambridge, MA: MIT Press.

Southern Poverty Law Center. "Criminal Justice Reform." Accessed June 2017. https://www.splcenter.org/issues/mass-incarceration

Southwick Associates. 2006. *The Relative Economic Contributions of U.S. Recreational and Commercial Fisheries*. Report. April 10. http://www.southwickassociates. com/wp-content/uploads/2011/10/Economics_Recreational_Commercial_ Fisheries_Harvests.pdf

Squires, Dale. 2010. "Fisheries Buybacks: A Review and Guidelines." *Fish and Fisheries* 11, no. 4: 366–87.

Squires, Dale, Harry Campbell, Stephen Cunningham, Christopher Dewees, R. Quentin Grafton, Samuel F. Herrick Jr, James Kirkley et al. 1998. "Individual transferable quotas in multispecies fisheries." *Marine Policy* 22 (2): 135–59.

Squires, Dale, and James Kirkley. 1999. "Skipper Skill and Panel Data in Fishing Industries." *Canadian Journal of Fisheries and Aquatic Sciences* 56, no. 11: 2011–18.

Strawson, Galen. 2015. "Against narrativity." In *Narrative, Philosophy and Life*, edited by Allen Speight, 11–31. Dordrecht: Springer.

Stuewer, Roger H. 1998. "History and Physics." *Science and Education* 7, no. 1: 13-30.

Tanana Chiefs Conference. 2017a. "Our Leadership." Tanana Chiefs Conference. Accessed June 2017. https://www.tananachiefs.org/about/our-leadership/

Tanana Chiefs Conference. 2017b. "Who We Are." Tanana Chiefs Conference. Accessed June 2017. https://www.tananachiefs.org/about/who-we-are/

Teh, Lydia C. L., and Ussif Rashid Sumaila. 2013. "Contribution of Marine Fisheries to Worldwide Employment." *Fish and Fisheries* 14, no. 1 (March): 77–88.

Tingley, Dustin, and Michael Tomz. 2014. "Conditional Cooperation and Climate Change." *Comparative Political Studies* 47, no. 3: 344–68.

Tobin, Paul. 2017. "Leaders and Laggards: Climate Policy Ambition in Developed States." *Global Environmental Politics* 17 (4): 28–47.

Toolan, Michael. 2012. *Narrative: A Critical Linguistic Introduction*. New York Routledge.

Tsebelis, George. 1989. "The Abuse of Probability in Political Analysis: The Robinson Crusoe Fallacy." *American Political Science Review* 83, no. 1: 77–91.

Tsebelis, George. 1990. *Nested Games: Rational Choice in Comparative Politics*. Berkeley: University of California Press.

Urban, Mark C. 2015. "Accelerating Extinction Risk from Climate Change." *Science* 348, no. 6234: 571–73.

Vanderheiden, Steve. 2008. *Atmospheric Justice: A Political Theory of Climate Change*. New York: Oxford University Press.

Vanderheiden, Steve. 2009. "Allocating Ecological Space." *Journal of Social Philosophy* 40, no. 2: 257–75.

Van Maanen, John. 1983. "The Boss: First-Line Supervision in an American Police Agency." In *Control in the Police Organization*, edited by Maurice Punch, 275–317. Cambridge: MIT Press.

Volden, Craig. 2005. "Intergovernmental Political Competition in American Federalism." *American Journal of Political Science* 49, no. 2: 327–42.

Ward, Artemus. 2004. "How One Mistake Leads to Another: On the Importance of Verification/Replication." *Political Analysis* 12, no. 2: 199–200.

Ware, Roger. 1984. "Sunk Costs and Strategic Commitment: A Proposed Three-Stage Equilibrium." *Economic Journal* 94, no. 374 (June): 370–78.

Waterman, Richard W., and Kenneth J. Meier. 1998. "Principal-Agent Models: An Expansion?" *Journal of Public Administration Research and Theory* 8, no. 2 (April): 173–202.

Weaver, Vesla M. 2007. "Frontlash: Race and the Development of Punitive Crime Policy." *Studies in American Political Development* 21, no. 2: 230–65.

Weaver, Vesla M., and Amy E. Lerman. 2010. "Political Consequences of the Carceral State." *American Political Science Review* 104, no. 4 (November): 817–33.

Wengraf, Tom. 2000. "Uncovering the General from the Particular. From Contingencies to Typologies in the Understanding of Cases." In *The Turn to Biographical Methods in Social Science: Comparative Issues and Examples*, edited by Prue Chamberlayne, Joanna Bornat, and Tom Wengraf, 140–63. London: Routledge.

Whelan, Eoin, and Robin Teigland. 2013. "Transactive Memory Systems as a Collective Filter for Mitigating Information Overload in Digitally Enabled Organizational Groups." *Information and Organization* 23, no. 3: 177–97.

Williams, J. Michael. 2009. "Legislating 'Tradition' in South Africa." *Journal of Southern African Studies* 35, no. 1: 191–209.

Williams, J. Michael. 2010. *Chieftaincy, the State, and Democracy: Political Legitimacy in Post-Apartheid South Africa*. Bloomington: Indiana University Press.

Williams, Kenneth. 2012. *Introduction to Game Theory: A Behavioral Approach*. New York: Oxford University Press.

Wilsnack, Richard W. 1976. "Explaining Collective Violence in Prisons: Problems and Possibilities." In *Prison Violence*, edited by Albert Kircidel Cohen, George F. Cole, and Robert G. Bailey, 61–78. Lexington, MA: Lexington Books.

Wilson, James Q., and Joan Petersilia. 2011. *Crime and Public Policy*. Oxford: Oxford University Press.

Woolf, Nicky. 2016. "Inside America's Biggest Prison Strike: 'The 13th Amendment Didn't End Slavery'." *Guardian*, October 22. https://www.theguardian.com/us-news/2016/oct/22/inside-us-prison-strike-labor-protest

World Ocean Review. 2017. "Catching Fish in International Waters." Accessed

April 2017. http://worldoceanreview.com/en/wor-2/fisheries/deep-sea-fishing/catching-fish-in-international-waters/

Yamane, David. 2000. "Narrative and Religious Experience." *Sociology of Religion* 61, no. 2: 171–89.

Yin, Robert K. 2011. *Applications of Case Study Research*. Thousand Oaks, CA: Sage.

Yin, Robert K. 2017. *Case Study Research and Applications: Design and Methods*. 6th ed. Thousand Oaks, CA: Sage.

Young, Kathryne M., and Joan Petersilia. 2016. "Keeping Track: Surveillance, Control, and the Expansion of the Carceral State." *Harvard Law Review* 129, no. 5: 1318–60.

Zagare, Frank C. 2009. "Explaining the 1914 War in Europe: An Analytic Narrative." *Journal of Theoretical Politics* 21, no. 1: 63–95.

Author Index

Subject Index

Note: Page numbers in *italics* indicate illustrated material.

abductive reasoning, 195, 201n10
accumulation of knowledge, 1, 5, 14, 17, 127, 192, 198
actions, specific to an information set, 60, *60*, 61
actors, criteria for inclusion, 59, *60*, 60–61
Afrobarometer survey, 75
agenda trees, 144–45
algorithm for narratives and models, 81
alternatives: multidimensional, 147; narratives, 155; packaged, 136, 137–38, *144*, 146–47, 150–51, 203n3
analytical narratives, 3, 41–42, 45, 52, 153, 199n5, 201n10
analytical transparency, 17
anthropology, 15
art of modeling, 196
assumptions: amending, 33, 34–35, 188, 193, 196; fewer the better, 64–65; fixed, 56; and generality of modeling, 63–65; mining, 35; realism in, 64, 67; role in model building, 24, 35, 56, 63–65; specificity of, 68; testing, 28; and theory, 22–24; weak assumptions, 63–64
assumptions-fitting approach, 35
awards and traditional identity, 88–89, 186–87
axiomatic transparency, 39, 186

backwards induction, 144, 162, 203n13. *See also* rolling back
Baron-Ferejohn bargaining model, 65
basic utility maximization, 59, 186
Bayes' rule, 170, 204n3
beliefs, *60*, 61
Bering Sea Pollock fishery, 96, 111
Borda count, 56
bycatch, 95–97, 103–6, 108–13, 115–19, 122–25, 202ch5n6, 202nn8–9, 203n11

The Calculus of Consent (Buchanan and Tullock), 55
carceral state, 157
cardinal utility, 54, *60*, 146
Centipede Game, 65
ceteris paribus clauses, 34
Chain Store Paradox, 165–66
chiefly authority, 81, 190
climate change, conclusion, 153; international diplomacy, 132–36; model design illustrated for, 129–53; narrative modeled as cooperative game, 146–52; overview, 129; social choice model, 136–46; story, 130–32; term usage, 133; testable propositions, 152–53; theory testing of model, 187–88; voting decisions, 143–46, *144*

233

theory-revisioning, 35
thick vs. thin narratives, 40–41
trade. *See* green trade/free green trade
trade gains, *148, 150, 151, 152*
traditional: authority, 74–76, 81, 88, 93;
community, 15, 83, 88, 89–91, 187;
governance, 76, 92; identity, 88–93;
leadership, 17, 72, 74–76
Tragedy of the Commons: application,
98–101, 105, 110, 117; and commercial fishing regulation, 98–99,
100, 103–6, 108; as extant knowledge,
99–102, *101, 102*; formal model of, 46;
and Prisoners' Dilemma (PD), 15, 99–
101, 189; resolution of, 103–4, 108;
theory-building model, 189
transitive preference, 52, 201ch3n1
transparency: analytic, 17; of assumptions, 1; axiomatic, 39, 186; clarity

and, 5, 28; in empirical research,
200n9; standard of, 14. *See also* rational choice theory

unanimity rule, 145, 152
utility function, 78–80, 85–87, 91–92,
121–23
utility maximization: Alaska fisheries, 99;
axiomatic transparency, 186; design
illustrated, 71–94

vending machines, models as, 68–69
voting decisions, 143–46, *144*

"will of the people" premise, 197
"World 3" (Popper), 42

YouTube, 132, 203nn1–2